Flavio Henrique Mendes

Trees falling from the road

Flavio Henrique Mendes

Trees falling from the road

Weather Effects and Microclimate Simulations

ScienciaScripts

Imprint
Any brand names and product names mentioned in this book are subject to trademark, brand or patent protection and are trademarks or registered trademarks of their respective holders. The use of brand names, product names, common names, trade names, product descriptions etc. even without a particular marking in this work is in no way to be construed to mean that such names may be regarded as unrestricted in respect of trademark and brand protection legislation and could thus be used by anyone.

Cover image: www.ingimage.com

This book is a translation from the original published under ISBN 978-3-330-76182-7.

Publisher:
Sciencia Scripts
is a trademark of
Dodo Books Indian Ocean Ltd. and OmniScriptum S.R.L publishing group

120 High Road, East Finchley, London, N2 9ED, United Kingdom
Str. Armeneasca 28/1, office 1, Chisinau MD-2012, Republic of Moldova, Europe
Managing Directors: Ieva Konstantinova, Victoria Ursu
info@omniscriptum.com

Printed at: see last page
ISBN: 978-620-8-37987-2

ACKNOWLEDGEMENTS

Thanks to Professor Demóstenes Ferreira da Silva Filho, (University of São Paulo), Jefferson Lordello Polizel, Computer Technician (University of São Paulo) and Professor António Manuel Saraiva Lopes (University of Lisbon);

Special thanks to the Piracicaba/SP/Brazil Fire Department, especially Douglas Gonçalves, for compiling and providing the database used, and also to the Piracicaba/SP Environmental Defence Department (SEDEMA), especially Juliana Gragnani, for providing information.

CONTENTS

SUMMARY

There are many benefits to the afforestation of public roads, such as increasing the useful life of asphalt through shading, filtering pollutants, absorbing CO2, intercepting rainwater and solar radiation and mitigating heat islands; however, for cultural reasons, it is often considered by the population as something negative, whose leaves clog gutters, roots destroy pavements, trunks and leaves get in the way of electrical wiring and, above all, they are susceptible to falls. The aim of this research was to study the behaviour of falling trees in the municipality of Piracicaba/SP, Brazil, with special attention to the wind regime in the city, based on data provided by the Piracicaba Fire Brigade, the Piracicaba Department for the Defence of the Environment (SEDEMA) and the Meteorological Station of the "Luiz de Queiroz" Higher School of Agriculture (ESALQ). Unprecedentedly, the main result was the direct relationship between falls and urbanised areas, which, with constant changes in land use, end up damaging the support of trees (roots), a fact justified by the high concentration of falls in the central part (36.7%), especially in the spring and summer seasons, i.e. wind and rain, which together accounted for 78.0% of the total. The expected return period for winds exceeding 75 km h'[1] , classified as storms on the Beaufort Wind Scale, was 2.8 events per year. While there was a predominance of south-easterly trade winds, belonging to the fourth quadrant, the direction of the wind gusts varied more and so, with a predominance of these winds, it is recommended that industrial parks be set up in the north, north-west and west zones, given the importance of preventing pollutants from entering the city. The most vulnerable species were: *Pachira aquatica* AubL (monguba), *Handroanthus* sp. (ipê roxo) and *Tipuana tipu* (Benth.) Kuntze (tipuana). By means of a computer microclimate simulation using the ENVI-met programme version 3.1, in two case studies, one in a residential neighbourhood and the other in José Bonifácio Square, it was possible to identify places where the most attention should be paid to falling trees, due to the roughness of the site, which can alter the speed and direction of the wind.

Keywords: ENVI-met version 3.1; Vulnerable species; Piracicaba/SP; Urban afforestation planning; Strong winds

CHAPTER 1

INTRODUCTION

Cities have been consolidated in a disorganised way, with no balance between green spaces and buildings, in which high levels of waterproofing modify the urban climate by favouring the occurrence of heat islands and the consequent reduction in people's quality of life (OKE, 1978).

The literature is vast in terms of the benefits provided by trees, but for cultural reasons, they are often damaged or eliminated to widen streets, repair pipes, maintain electrical wiring, build or renovate houses (LIMA, 1993), whose urbanisation can directly interfere with falls (MENDES et aL, 2015).

Milano (1996) explained at the Brazilian Urban Afforestation Congress (CBAU) in Salvador/BA that success in urban afforestation consists of five stages: planning, executing, controlling, analysing and replanning. The city of São Paulo, for example, has had fallen trees mapped since 2003, and this is an important tool for management.

The general objective of the research was to assess the behaviour of falling trees in Piracicaba/SP, while the specific objectives were: (i) to identify regions where falls are concentrated; (ii) the influence of climatic variables and (iii) species vulnerability.

As this is an unprecedented study that can be applied to other cities, Chapter 2 provides a broad overview of the effect of climatic variables on tree falls in Piracicaba/SP, with special emphasis on wind. Chapter 3 studied the leaf area index of common species found in local urban tree planting, using technological equipment (LAI 2000 *Plant Canopy Anatysefj.* This chapter served as the basis for the next, 4 and 5, which were two case studies, respectively in a residential neighbourhood and in the city's central square, José Bonifácio, using microclimate simulations with ENVI-met v.3.1 *software,* in order to understand the area's behaviour under predefined conditions.

References

LIMA, A.M.L.P. **Piracicaba/SP:** Analysis of roadside afforestation in the central area and its surroundings. 1993. 283p. Thesis (Doctorate in Plant Science) - Escola Superior de Agricultura "Luiz de Queiroz", Universidade de São Paulo, Piracicaba, 1993.

MENDES, F.H.; ZANLORENZI, H.C.P.; DOBBERT, L.Y.; PRATA-SHIMOMURA, A.R.; SILVA FILHO, D.F. Microclimatic changes in José Bonifácio Square in Piracicaba (SP). In: CONGRESSO BRASILEIRO DE ARBORIZAÇÃO URBANA, 19., 2015, São Paulo. **Proceedings...** São Paulo: CBAU, 2015. 11p.

MILANO, M.S. Urban afforestation in Brazil: myths and reality. In: CONGRESSO BRASILEIRO DE ARBORIZAÇÃO URBANA, 3., 1996, Salvador. **Proceedings...** Salvador: CBAU, 1996. p. 1-6.

OKE, T.R. **Boundary layer climates.** London: Methuen, 1978. 372p.

CHAPTER 2

THE EFFECT OF WIND ON FALLING TREES IN PIRACICABA/SP

Summary

The aim of this research was to study the behaviour of falling trees in the municipality of Piracicaba/SP, Brazil, with special attention to the local wind regime. The Fire Department provided a database of falls containing 275 cases from 3 January 2011 to 24 May 2014, with their respective locations, which were spatialised using GIS. The effect of urbanisation played a major role in falls, with the central region concentrating most of them (36.7%). The spring and summer seasons accounted for 78.0 per cent of the falls, which correspond, respectively, to the periods of most intense winds and rains. The expected return period for winds above 75 km h'^{1} , classified as storms on the Beaufort wind scale, was 2.8 events per year. The species most vulnerable to falling were *Pachira aquatica* AubL (monguba), *Handroanthus* sp. (ipê roxo) and *Tipuana tipu* (Benth.) Kuntze (tipuana), characterising specific management opportunities. While there was a predominance of south-easterly trade winds, belonging to the fourth quadrant, the direction of the wind gusts varied more and so, with a predominance of south-easterly winds, it is recommended that industrial parks be set up in the north, north-west and west, given the importance of preventing pollutants from entering the city.

Keywords: Wind regime; Urbanisation; Vulnerability of species variation and, thus, since the predominance of Southeast winds, it is recommended to spot industrial parks at North, Northwest and West zones, given the importance to prevent pollutants entering the city.

Keywords: Wind regime; Urbanisation; Vulnerability of species

1.1 Introduction

There are many benefits to tree planting on public roads (SHASHUA-BAR; HOFFMAN, 2000; DOBBERT et aL, 2014). However, trees can be vulnerable to being knocked over and falling, with wind being one of the aggravating agents (MATTHECK; BRELOER, 1997).

According to the Aurélio Dictionary, "vulnerable" is understood as "the weak side of an issue or the point where someone can be hurt or attacked". Julião et aL (2009) define vulnerability as the "degree of loss of an exposed element or set of elements, as a result of the occurrence of a natural, technological or mixed process (or action) of certain severities".

Wind is caused by differences in atmospheric pressure, with air moving from areas of higher pressure to those of lower pressure, creating different speeds, as well as the Coriolis force, which arises due to the Earth's rotational movement, and also centripetal and friction forces (BONAN, 2002).

There are three types of atmospheric currents on the globe: (i) trade winds, between latitudes 30° N and 30° S, which are weak in intensity; (ii) westerly winds, between 30° and 60° N and S, which are fast and (iii) polar winds, at latitudes greater than 60°, which are stormy. In Brazil, the southeast trade winds predominate (PRATA, 2005), as shown in figure 2.1.

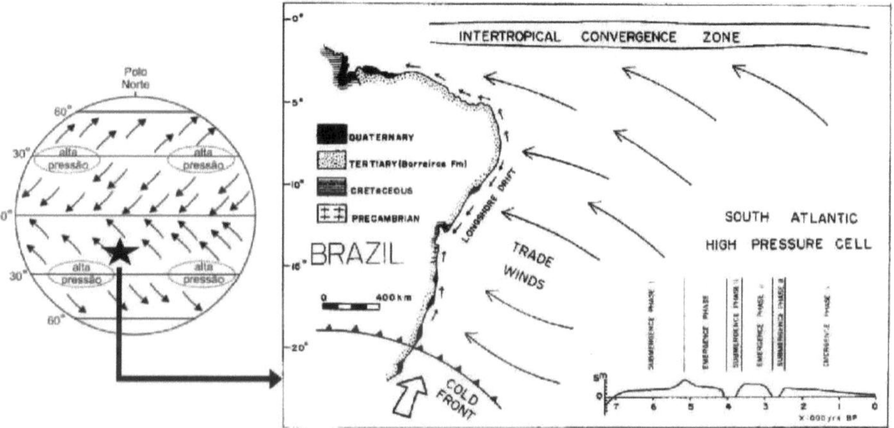

Figure 2.1 - Earth's wind regime, with emphasis on the south-east trade winds. Source: Adapted from Dominguez, Bittencourt and Martin (1992)

Predicting the weather in Piracicaba/SP with accuracy is complicated by the flow of air masses throughout the year. In the southeast, there is a predominance of the Atlantic Tropical Mass (mTa) during the summer, characterised as hot and humid, originating in the South Atlantic, while in winter there is the Atlantic Polar Mass (mPa), cold and humid, originating in Antarctica (Figure 2.2).

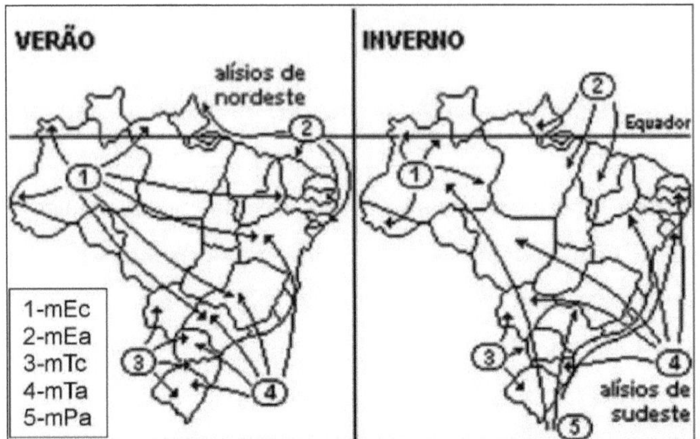

Figure 2.2 - Influences of air masses on Brazilian territory. Source: Adapted from Sene and Moreira (1998)

In urban environments, the wind is responsible for dispersing pollutants and controlling the temperature. Research carried out by the Institute of Astronomy, Geophysics and Atmospheric Sciences at the University of São Paulo (IAG/USP) has proven that the São Paulo Metropolitan Region (the capital and 38 other neighbouring municipalities) is a pollutant exporting centre: 20 to 30% of the pollution found in Campinas, Tatuí and Sorocaba comes from São Paulo, due to the wind regime which, depending on its intensity, can reach Bauru/SP, 400 km from the São Paulo capital. What mitigates the situation in São Paulo is the sea breeze from the Atlantic Ocean, which circulates 500 metres above the surface, reducing the temperature and dispersing pollutants. On the other hand, Cubatão, at the foot of the Serra do Mar, is directly responsible for the ozone found on the São Paulo plateau (FIORAVANTI, 2002).

At the micro-scale, the wind can be altered by the urbanisation process through the roughness of areas (CLEUGH; GRIMMOND, 2011), which is a surface friction value, i.e. heights of built-up masses capable of significantly altering the movement of air masses (SANTOS, 2004).

Below are approximate values for different surfaces from the European Wind Atlas (TROEN; PETERSEN, 1989) (Table 2.1).

Table 2.1 - Roughness values (Z0) for different surfaces

zo	**Terrain characteristics**
1,00	City
0,80	Forest
0,50	Suburbs
0,30	Tree belts
0,20	Trees and shrubs
0,10	Farm with closed vegetation
0,05	Doing it with open vegetation
0,03	Farm with few trees/buildings
0,02	Airport areas with buildings and trees
0,01	Airport runway areas
0,008	Pasture
0,005	Ploughed soil
0,001	Snow
0,0003	Sand
0,0001	Water (lakes, rivers and oceans)

Source: Adapted from Troen and Petersen (1998)

Figure 2.3 shows the behaviour of the wind according to different typologies (urban centre, suburb and open countryside).

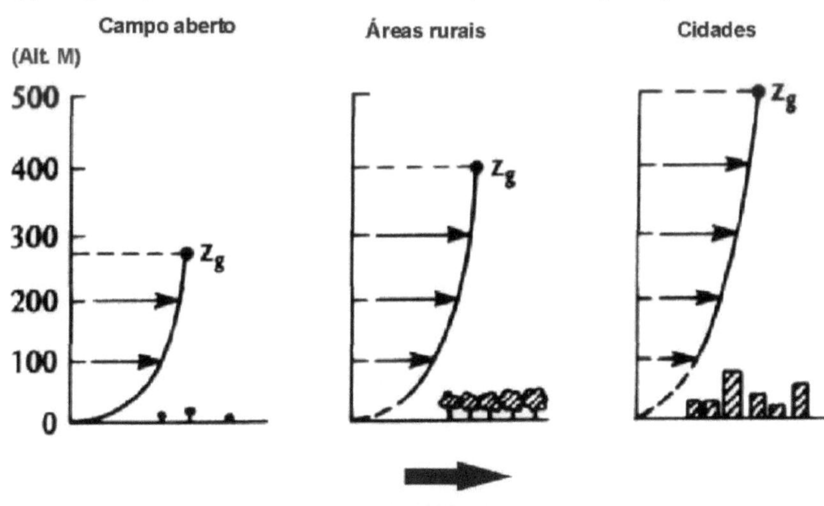

Figure 2.3 - Decrease in wind speed influenced by variation in terrain roughness. Source: Adapted from Oke (1987)

For falling trees, Albers et al. (1992) assesses that there are seven categories responsible: decline, cracks, root problems, weak branch union, cankers, poor architecture and dead wood. In order to guarantee the integrity of the roots, the authors propose the concept of *Criticai rootradius* (CRR), which is represented by the ratio DBH x 1.5 and if more than 40 per cent of the area is damaged, stability is compromised (Figure 2.4).

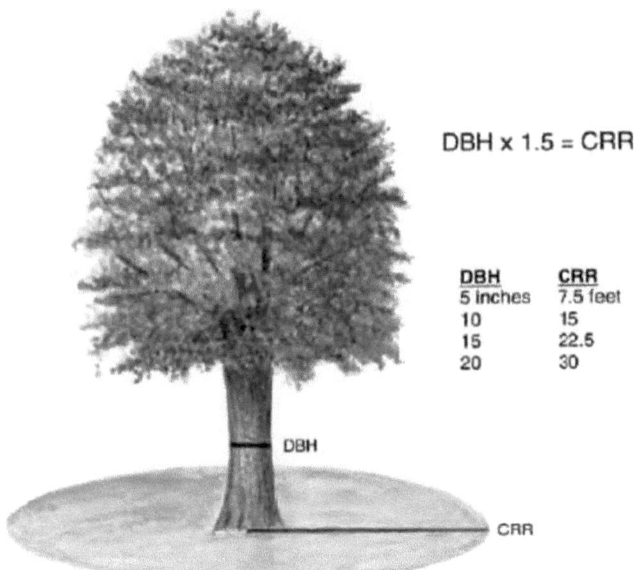

Figure 2.4 - Representation of the critical area for preserving root integrity.
Source: Albers et al. (1992)

Seitz (1995) directly correlates forest species and wood degrading fungi responsible for breakage, whose inclement weather (rain, wind, etc.) intensifies the process, except in sheltered conditions and dry environments, where branches can remain connected for years. Green pruning capable of unbalancing the crown should be avoided, as the epicormic shoots (unconnected to the central part of the tree) are fragile to breakage.

Mattheck and Breloer (1997), experts in biomechanics, reported on the stresses to which trees are exposed, such as: traction, compression, shear and bending and twisting moments. In order to avoid toppling over, they create internal tensions that alter their shape, size and structure, an event known as adapted growth.

Mattheck (2007) evaluated the *slenderness* coefficient as another critical factor for tree fall. This index can be obtained by dividing height (m) by diameter (cm), i.e. H/D, and if it is greater than 50, the tree is vulnerable, as are internal cavities greater than two thirds (Figure 2.5).

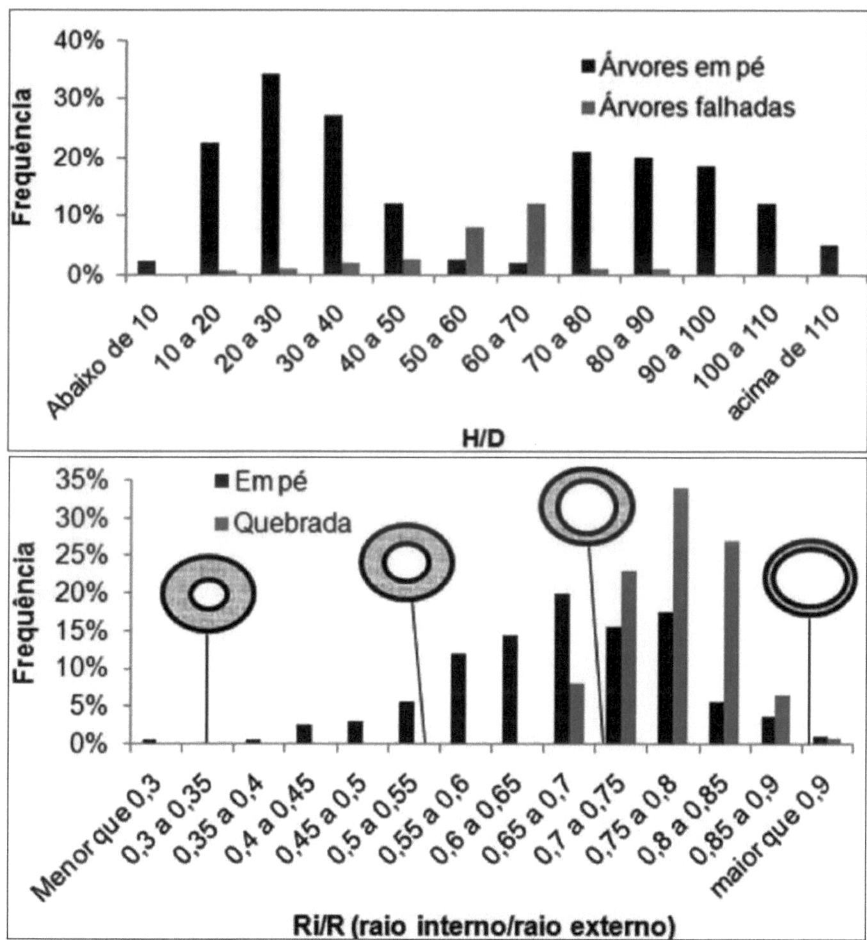

Figure 2.5 - Reported biomechanical weaknesses in trees, with the slenderness coefficient and the critical amount of hollow. Source: Adapted from Mattheck (2007)

The aim was to assess the behaviour of falling trees in the municipality of Piracicaba/SP.

2.2 Materials and Methods

2.2.1 Study area

The research was based on the urban boundaries of the municipality of Piracicaba/SP, made available by IPPLAP (Piracicaba Research and Planning Institute) in 2010. The city is 164 km from the state capital, with geographical coordinates of 22°42'30" S and 47°38'00" W, a Cwa type climate, according to the Kõppen classification, at an altitude of 546 m and an annual rainfall of 1328 mm. Its predominant vegetation is stationary semi-deciduous forest, located in the domain of the Paulista Peripheral Depression, a region lowered by erosion, between the highlands of the Atlantic Plateau and the high escarpments of the basaltic cuestas of the Western Plateau (OLIVEIRA; PRADO, 1989).

2.2.2 Database

An organised database of tree falls from 3 January 2011 to 24 May 2014 was obtained from the 16th Fire Brigade of Piracicaba/SP, accounting for 3.5 years of returns, with 275 cases, which contained the address and date/time of the call from the citizen.

The methodology has some limitations:

(i) Data prior to that date, printed on paper and not compiled, ended up being lost over time and could not be retrieved; today the database is computerised;

(ii) Falls that posed a risk to people or public property were counted;

(iii) The database did not separate falling trees from falling branches;

(iv) The time of registration corresponds to the time when the vehicle leaves (answering the call) and not exactly the same time as the crash;

(v) SEDEMA (the Department for the Defence of the Environment) also has data on tree falls (probably more than the Fire Brigade), but without specific data, such as the time of occurrences; however, the Department has made efforts to improve this. In more urgent weather events, both organisations join forces for the same cause, with mutual support, and can even offer support to towns in the region.

2.2.3 Spatialisation of points

Once we had the falls with their respective addresses (street, number, municipality - state), they were spatialised in GIS using the BatchGeo LLC *website*[1] , the free version of which allows up to 250 records to be spatialised at a time, so this procedure had to be done in two parts. As not all of the falls fell exactly at the right point, they were checked one by one (using *Google Street View,* image 2011) with the respective adjustments so that the 275 falls could then be merged into a single *shapefile* file.

2.2.4 Assumption of species

Also using the *Google Street View* tool, alternating with the *"go2streetvievV" plugin,* installed in the Quantum GIS programme version 2.6, species were assumed, where possible (Figure 2.6). The use of this methodology in the area of urban afforestation is unprecedented. It is worth noting that this was possible due to the conciliation between the Fire Department's falls (from January 2011) and *Google*'s street level image for the municipality (April 2011 and then August 2015), i.e. through images before and after the fall. In some cases it was not possible to identify the species, such as falls between January and April 2011 or because there was more than one tree in front of the address, such as large frontages, outward-facing walls or places with green areas in front.

---

1 <http://pt.batchgeo.com/>. Accessed on: 12 August 2015.

Figure 2.6 - Fallen tree of the sibipiruna species *(Poincianella pluviosa* (DC.) L.P. Queiroz), on the right, spatialised and assumed, located at Rua Floriano Peixoto, 799, in the Centro neighbourhood. Source: Google Street View

2.2.5 Climate variables

Correlations with climatic variables were based on data from the ESALQ Meteorological Station[2] , of which: (i) Conventional Station, operating since 1917, but only in 1943 with measurements of wind, relative humidity and atmospheric pressure and (ii) Automatic Station, since 1997, with sensors for precipitation, temperature, humidity, radiation, evapotranspiration, wind speed and direction.

Climatological normals are defined every 30 years, starting in 1901 and

[2] <http://www.leb.esalq.usp.br/posto/>. Accessed on: 25 September 2015.

so on. However, when the period is not complete and there are at least 10 years, Provisional Climatological Normals can be made, as explained in Technical Document WMO-TD/No. 341 of the World Meteorological Organisation (WMO, 1989).

To study the direction and speed of the prevailing wind, we used data from the Automatic Station (1997 to 2014), at a height of 2 metres, recorded every 15 minutes, while the maximum wind gusts and precipitation, daily data from the Conventional Station, at a height of 10 metres.

It is important to note that it was sometimes necessary to correct to UTC - 3 due to daylight saving time, as the station disregarded the event.

Since the station's data is available in Julian Days (1 to 365 or 366), the *website* proposed by Teimo Ghiorzi[3] was used to convert it to the day/month/year calendar. The working scheme that summarises the methodology is shown in figure 2.7.

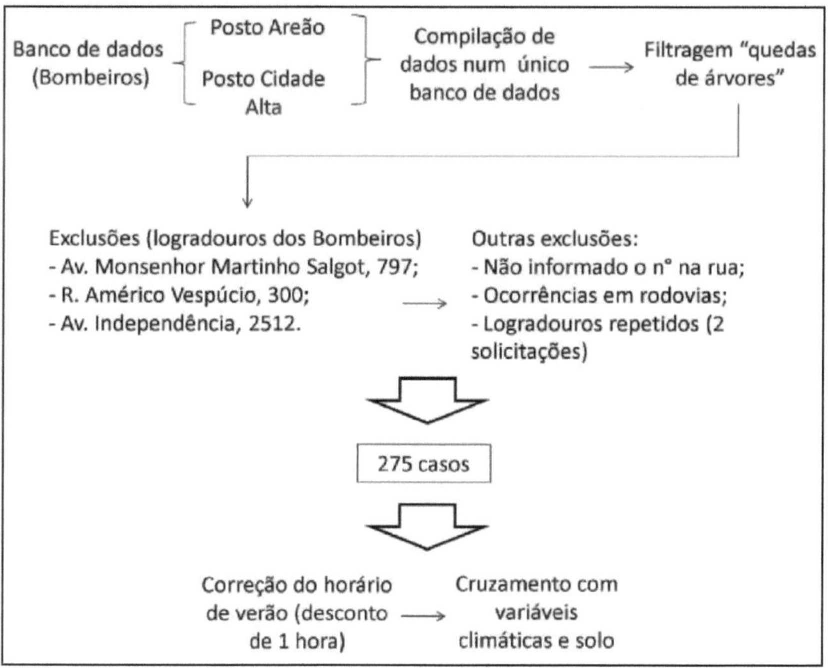

[3] <http://ghiorzi.org/diasjuli.htm>. Accessed on: 25 June 2015.

Figure 2.7 - Work scheme used to conduct the research, from database acquisition, filtering and cross-referencing with climate variables and soil type. Source: the author

2.3Results and Discussion

2.3. 1Wind

A gale of 15 m s^{-1} (54 km h^{-1}) is normally the value used as a reference by insurers in property damage claims (CAMARGO, 2014), and the direction of the prevailing wind may or may not coincide with the direction of the gust (BURGARDT; LEITE; VIRGENS FILHO, 2011), and the tree may fall depending on the path through which the gale passes.

Firstly, the wind speed correlation between the two ESALQ weather stations, the Automatic (2 m) and the Conventional (10 m), was analysed for the 275 cases of falls. The coefficient of determination (R^2 = 0.7314) showed a linear trend in the data (Figure 2.8), while the Pearson correlation coefficient, which ranges from -1 to 1, where the extremes represent a strong correlation, being negative or positive respectively, was 0.86, classified as strong (DANCEY; REI DY, 2006).

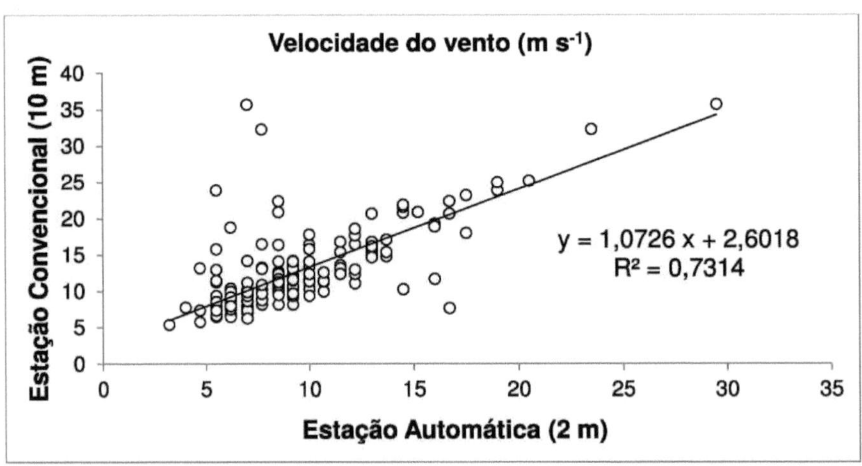

Figure 2.8 - Strong correlation between wind speeds measured at ESALQ's meteorological stations. Source: the author

The work by Oliveira and Lopes (2007) found that trees start to fall in winds of

7 m s^{-1} (at a height of 10 m); in this case, 97.1% behaved in the same way. For speeds above 10 m s^{-1} (36 km tr^{-1}), 81.8% and, according to insurance claims (above 15 m s^{-1}), 42.2% of the total.

Figure 2.9 shows the prevailing wind direction in the region, alluding to the south-east trade winds, according to the Automatic Station, since the Conventional Station did not have this data. It can be seen that the main wind directions belong to the 4° quadrant (east, southeast and south), and it can be inferred that the main regions in the municipality for allocating industrial districts are on the opposite side, i.e. north, northwest and west, so that pollutants don't enter the city.

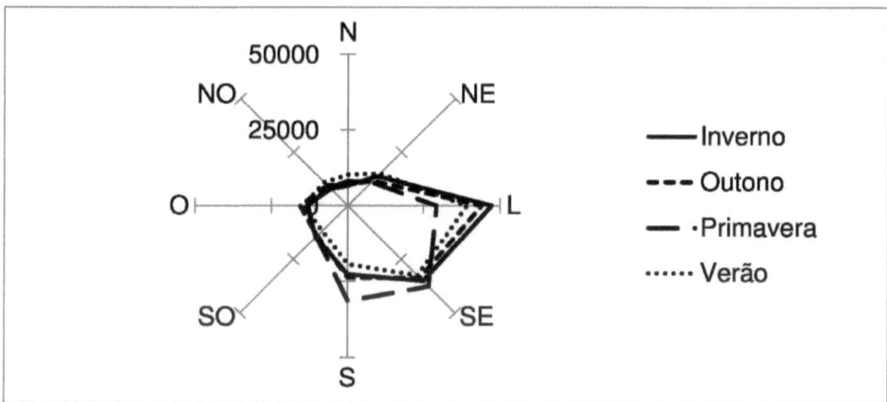

Figure 2.9 - Prevailing wind direction during the four seasons. Source: the author

Figure 2.10 shows that spring is the windiest season of the year, but it is in summer that the highest gusts and thunderstorms occur, with average highs of 11.2 m s .$^{-1}$

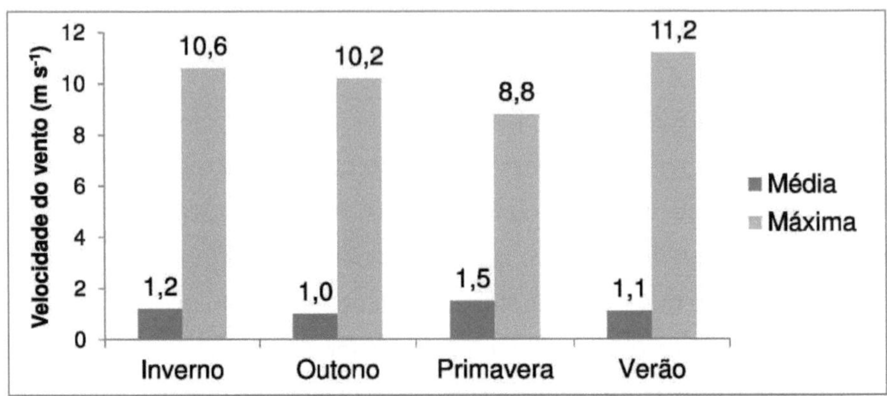

Figura 2.10 - Wind speed: average of the average and average of the maximum of the Automatic Station during the four seasons of the year. Source: the author

Figure 2.11 alludes to figure 2.10, broken down into directions according to the gusts of wind every 15 minutes. It is clear that the direction associated with the gusts varied more throughout the year, compared to the predominance of south-easterly trade winds in the city (quadrant 4).

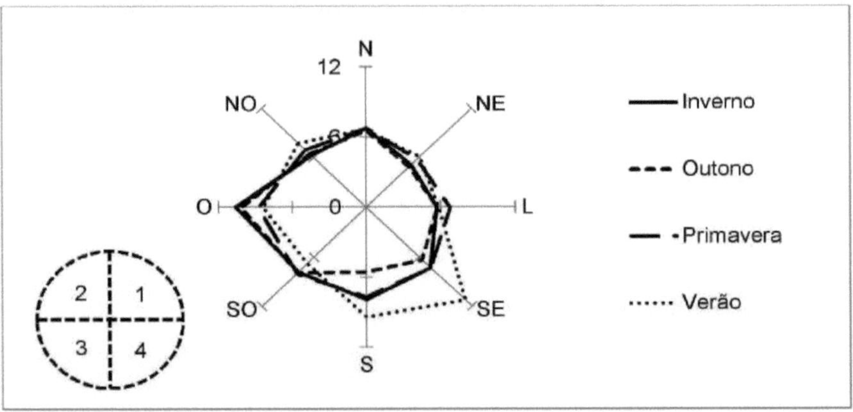

Figure 2.11 - Direction of wind gusts throughout the seasons, with the respective average maximum speed (m s^{-1}); in the left-hand corner there is the standard numbering of the 4 quadrants. Source: the author

Based on the Beaufort wind scale, which helps to interpret the maximum wind speed data (gusts), the expected return period for each corresponding band is given (Table 2.2). The data was compiled from the Conventional

Station (1964 to 2014).

Table 2.2 - Beaufort wind scale and respective expected return period, in days

Grade	Description	Speed (km h)$^{-1}$	Return period (days)
0	Calm	0-2	1087,3
1	Calm Wind	2-6	101,0
2	Mellow Breeze	7-11	47,8
3	Light breeze	12-19	7,7
4	Moderate Breeze	20-29	3,2
5	Strong Breeze	30-39	3,4
6	Strong wind	40-50	6,2
7	Very Strong Wind	51 -61	25,2
8	Very strong wind	62-74	55,0
9	Temporal	75-87	188,5
10	Strong storm	88-101	565,4
11	Very Strong Thunderstorm	102-117	2355,8
12	Taken, Hurricane	>118	3533,8

Source: the author

It was found that the most frequent winds fell within the range of 20 to 39 km h^{-1} , i.e. 5.5 to 10.8 m s^{-1} . For comparison purposes with insurance claims (15 m s^{-1}), the expected return period for this speed is 1 every 19 days. For natural phenomena classified as "thunderstorm" or "tornado/hurricane" (over 75 km h^{-1}), for example, the expected return period is 1 every 128.5 days, i.e. 2.8 events per year. The main highlight was 29 March 2006, when the equipment reached its maximum measurement value (44 m s).$^{-1}$

It is difficult to establish correlations with the prevailing wind direction for the trees that have fallen, since, as mentioned in the methodology, the time of recording corresponds to the time of the call and not exactly to the time of the fall.

The influence of rainfall on the waterlogging of the soil and the consequent fall of the tree should not be overlooked: 65.5 per cent of falls occurred during periods of rain in the last 24 hours (Figure 2.12), with 47.6 per cent falling during periods with at least 5 mm of rain up to a day before.

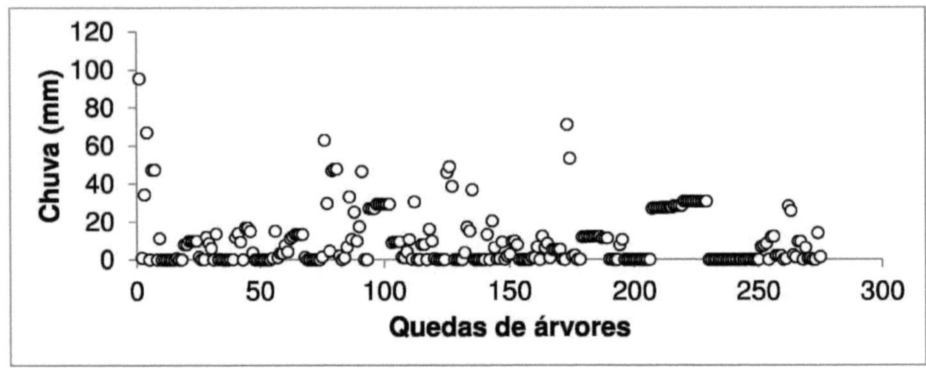

Figure 2.12 - The influence of rainfall on tree falls is another aggravating factor, in addition to wind speed. Source: the author

2.3.2 Places of greatest occurrence

According to the city's five geographical regions, the central part showed the most falls, 36.7%, followed respectively by the east (22.2%), north (20.0%), south (13.1%) and west (8.0%). For Mendes, Silva Filho and Lopes (2015), the shadows of buildings can be considered indicative of the degree of urbanisation in the area. Thus, the centre was the region with the most shadows (WorldView-2 satellite image, 22 April 2011), highlighting the "urbanisation" factor as one of the main aggravating factors in tree falls (Figure 2.13), in which damaged roots, compacted soil and narrow flowerbeds are common problems.

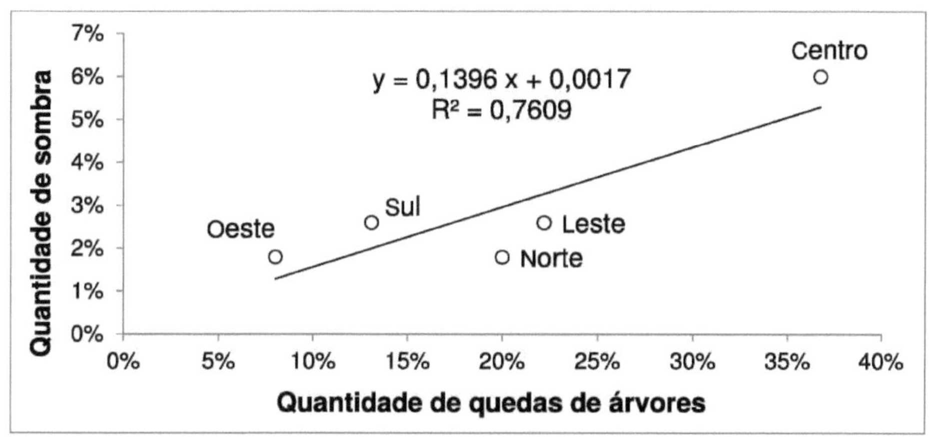

Figure 2.13 - Correlation between tree falls and the degree of urbanisation, represented by the amount of shade from buildings. Source: the author

Cross-referencing the information on fallen trees with the type of soil, which comes from the IAC (Agronomic Institute of Campinas), once again revealed that the "urbanisation" factor influenced the falls, accounting for 48.4% of the falls, in an area of the territory quantified at 29.3% (Figure 2.14).

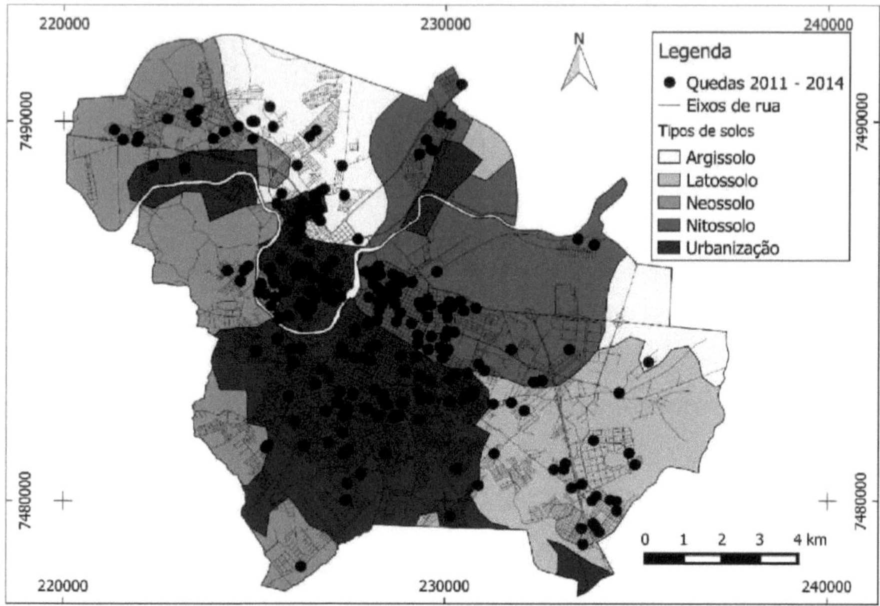

Figure 2.14 - Spatialisation of tree falls recorded by the Piracicaba Fire

Department, according to soil type from the Campinas Agronomic Institute (IAC). Source: the author

The neosol, present in large quantities in the municipality of Piracicaba/SP (OLIVEIRA et aL, 1999), has an A horizon less than 40 cm thick, deposited directly on the rock or C horizon (soil and rock fragments). Characterised by problems of drought in winter and waterlogging in summer, the sloping terrain makes it difficult for the plant to take root.

The falls were concentrated in the spring and summer seasons, characterised by wind and rain, in that order, which accounted for 78.0%. Specifically for this analysis, the period from 1 January 2011 to 31 December 2013 was used, i.e. exactly 3 full years, totalling 250 falls, to standardise the number of days in each season (Figure 2.15).

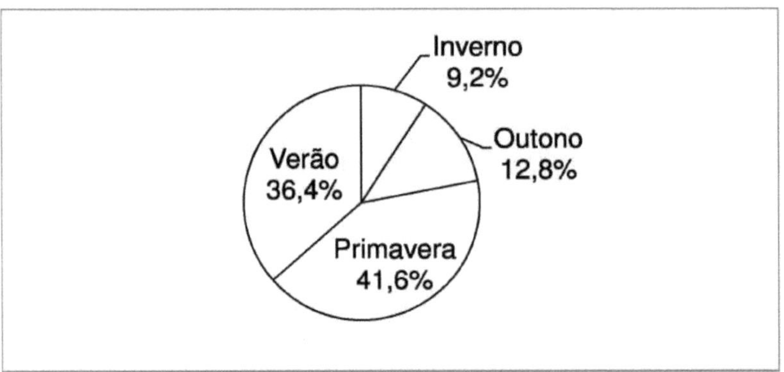

Figure 2.15 - Fallen trees by season. Source: the author

Ribeiro and Lopes (2011) found that streets orientated N-S recorded higher numbers of falls in Lisbon, the capital of Portugal, due to the northerly wind, known as "nortada", which is common in summer. Due to the urban planning of the roadbeds, together with the direction of the prevailing south-easterly trade winds in the area, it can be seen that the streets with a south-east - north-west (SE-NO) and north-east - south-west (NE-SO) orientation accounted for 80.4% of the falls (Figure 2.16).

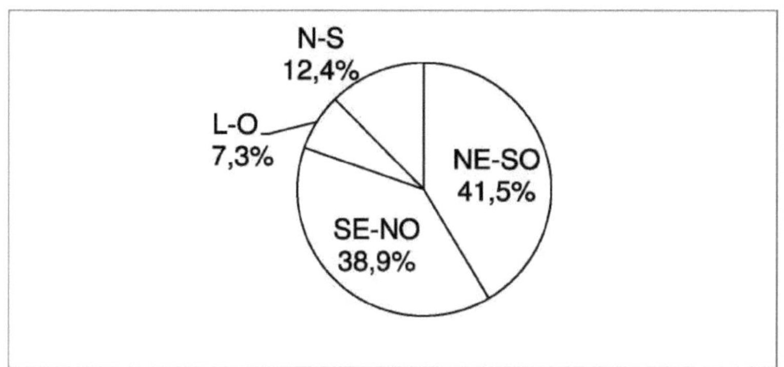

Figure 2.16 - Fallen trees by street orientation. Source: the author

The division of the north, south, east, west and centre geographical regions was planned appropriately, following the topography of the relief, whose territorial limits correspond to hilltops and lowlands. The digital terrain model, made with 1 m contour lines obtained from Topodata, under the initiative of INPE (National Institute for Space Research), revealed that the falls in the São Judas / São Dimas neighbourhoods are susceptible to higher intensity winds, due to the higher altitude of the terrain (Figure 2.17).

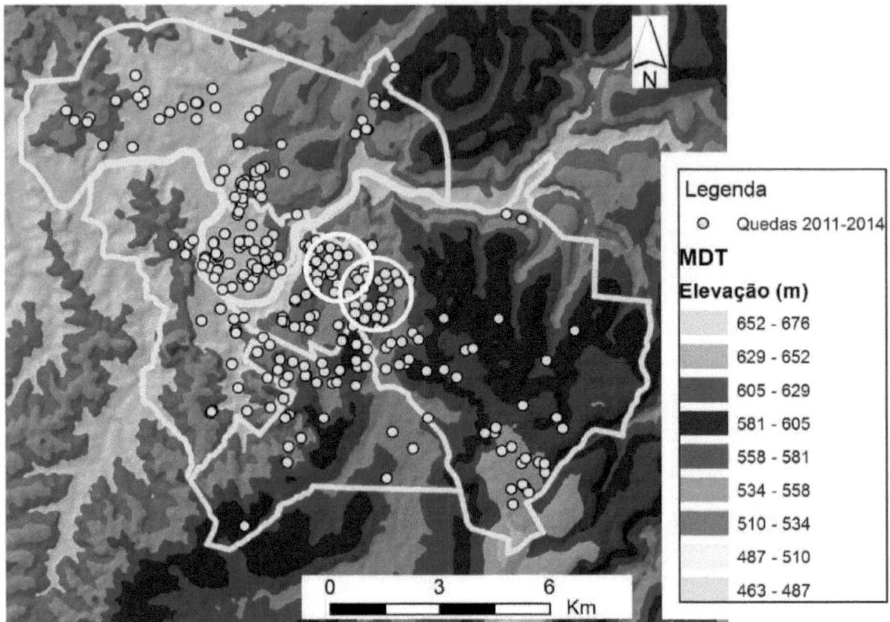

Figure 2.17 - Fallen trees according to the digital terrain model (DTM) of Piracicaba/SP; the white circle includes the São Judas and São Dimas neighbourhoods, at high altitude. Source: the author

2.3. 3Species vulnerability

A total of 123 individuals were assumed, as well as 4 dead trees. Rollo (2014) sampled 9.6 per cent of Piracicaba's 3759 blocks, or 360, and found 5744 individuals, in a tree and shrub population estimated at 60,146 street individuals in the city, belonging to 165 species.

In order to be representative of the fall assumptions, at least 2 cases were considered and compared with the results of Rollo (2014), in order to check whether those who fall more are because there are more of them or whether they are in fact more vulnerable. The index was obtained by dividing the assumed number of falls by the number on the streets and transformed into a percentage (Figure 2.18).

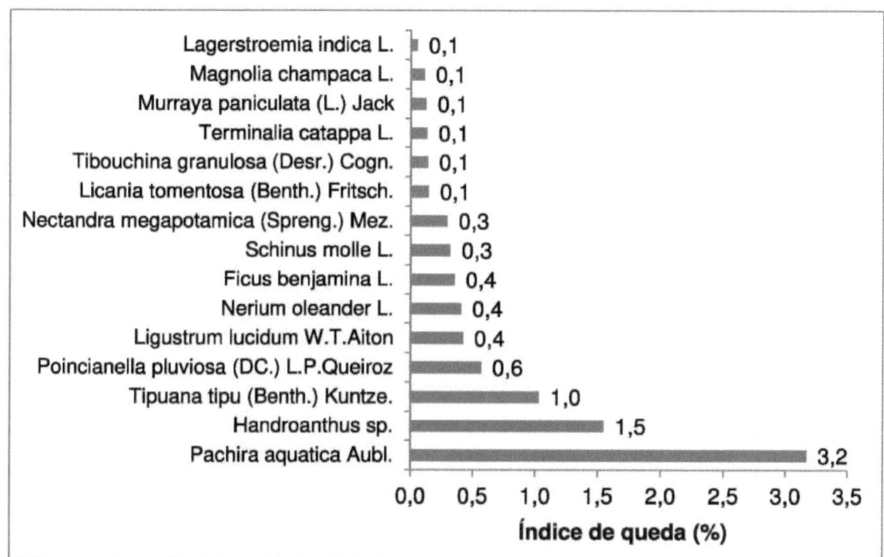

Figure 2.18 - Species planted in street trees that are vulnerable to falling, according to the 275 cases in the Fire Brigade database over the past 3.5 years. Source: the author

The most vulnerable species were *Pachira aquatica* AubL (monguba), *Handroanthus* sp. (ipê roxo) and *Tipuana tipu* (Benth.) Kuntze (tipuana), as shown in figure 2.19.

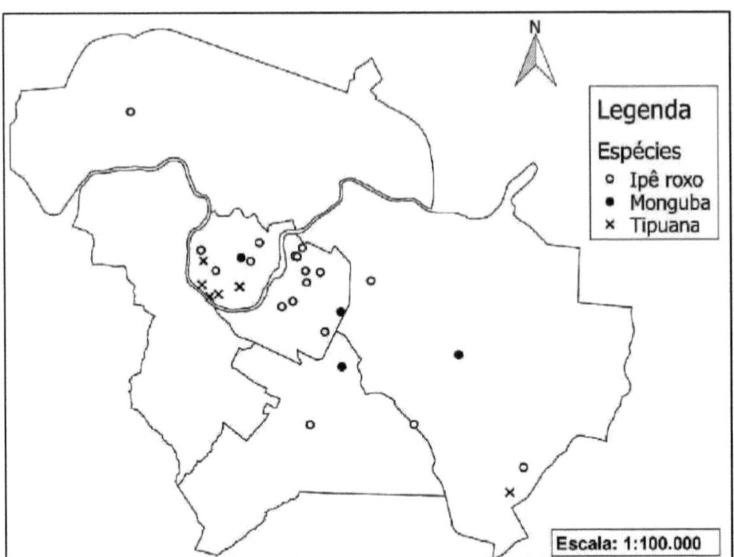

Figure 2.19 - Spatialisation of the three species most vulnerable to falling: monguba, purple ipê and tipuana, with 71.4% of cases concentrated in the central region. Source: the author

Despite only visual analysis, the species *Schizolobium parahyba* (Vell.) S.F. Blake (guapuruvu) and *Triplaris americana* L. (antwood), rare in streets but common in squares, often fail at the trunk when twisted. Both those responsible for removing the fallen trees, the Fire Brigade and SEDEMA, showed a satisfactory level of knowledge on the subject, since representatives from the Fire Brigade said that tipuana, sibipiruna and ipê fall a lot, while representatives from SEDEMA said ipê and ficus benjamina.

Figure 2.20 highlights the importance of the size of the flowerbed, since if it is narrow, the roots tend to rise to breathe and compromise support, making them vulnerable to falling, as observed in Rua Pedro Eusébio Stocco, 362 (Jardim Alvorada neighbourhood) and Rua dos Uirapurús, 133 (Parque Chapadão neighbourhood), respectively.

Figure 2.20 - Examples of plants with small beds that have fallen over: *Neríum oleanderL.* (*spirea*) and *Licania tomentosa* (Benth.) Fritsch. (oiti), in that order. Source: Google Street View

2. 4Conclusion

The research contributed to an understanding of tree falls in the municipality of Piracicaba/SP, with the "urbanisation" factor being one of the main culprits, since 36.7% were concentrated in the downtown area, while 48.4% were in soil classified as urbanised. This is due to stressful conditions for the plants, such as compacted soil, narrow beds and damage to the root system, which can jeopardise support.

The seasons in which the most trees fell were spring and summer, which together accounted for 78.0 per cent of the total. This is precisely the period when the aggravating factors of rain (summer) and wind (spring) occur and, depending on the phytosanitary conditions, can cause trees to fall. While 97.1 per cent of the trees fell when the wind gust exceeded 7 m s^{-1} in the last 24 hours, according to the methodology proposed by Oliveira and Lopes (2007), almost half (42.2 per cent) fell when the wind gust exceeded 15 m s^{-1} . Of the total, 65.5 per cent of the trees fell when it rained in the last 24 hours.

The expected return period for strong winds, classified as "thunderstorm" or "tornado/hurricane" according to the Beaufort wind scale, with a speed of over 75 km h^{1} , is 1 every 128.5 days, i.e. 2.8 events per year, while winds over 100 km tr^{1} , 1 every 1087 days (equivalent to 3 years).

The most vulnerable species in the conditions analysed were *Pachira aquatica* AubL (monguba), *Handroanthus* sp. (ipê roxo) and *Tipuana tipu* (Benth.) Kuntze (tipuana), which represent specific management opportunities where the intrinsic conditions of the planting site may be unfavourable.

Based on the direction of the prevailing easterly, south-easterly and southerly winds (trade winds), it was concluded that the best places to locate industries are in the north, north-west and west, given the importance of dispersing pollutants, preventing them from entering the interior of cities.

References

ALBERS, J.S.; POKORNY, J.D.; JOHNSON, G.R. How to Detect and Assess Hazardous Defects in Trees. In: POKORNY, J.D. (Ed.). **Urban Tree Risk Management:** A Community Guide to Programme Design and Implementation. St. Paul: USDA Forest Service. 1992. chap. 3. p. 41-116.

BONAN, G.G. **Ecological Climatology:** Concepts and Application. New York: Cambridge Univ. Press, 2002. 990p.

BURGARDT, S.; LEITE, M.L.; VIRGENS FILHO, J.S. Prevailing hourly and gust wind direction in locations in and around Campos Gerais, Paraná. **Revista Brasileira de Climatologia**, Curitiba, v. 8, p. 137-151,2011.

CAMARGO, R. **Boletim Climatológico Anual da Estação Meteorológica do IAG / USP / Secção Técnica de Serviços Meteorológicos**, São Paulo, v. 17, 2014. 76p.

CLEUGH, H.; GRIMMOND, S. Urban Climates and Global Climate Change. In: HENDERSON-SELLERS, A.; MCGUFFIE, K. **The Future of the World's Climate,** Oxford: Elsevier. 2011. chap. 3. p. 47-76.

DANCEY, C.; REIDY, J. **Statistics Without Maths for Psychology:** Using SPSS for Windows. 3.ed. Porto Alegre: Artmed, 2006. 608p.

DOBBERT, L.Y.; PRATA-SHIMOMURA, A.R.; MENDES, F.H.; SILVA FILHO, D.F. The influence of tree canopy cover on urban thermal comfort. In: CONGRESSO LUSO - BRASILEIRO PARA O PLANEAMENTO URBAN REGIONAL INTEGRADO E SUSTENTÁVEL, 6., 2014, Lisboa. **Proceedings...** Lisbon: PLURIS, 2014. 8p.

DOMINGUEZ, J.M.L.; BITTENCOURT, A.C.S.P.; MARTIN, L. Controls on Quaternary Coastal evolution of the east-northeastern coast of Brazil: roles of sea-level history, trade winds and climate. **Sedimentary Geology**, Amsterdam, v. 80, n. 3-4, p. 213-232, 1992.

FIORAVANTI, C. **Greenhouse that exports pollution.** Pesquisa FAPESP. 2002. p. 28-35.

JULIÃO, R.P.; NERY, F.; RIBEIRO, J.L; CASTELO BRANCO, M.; ZÊZERE, J.L. Methodological guide for the production of municipal risk cartography and for the creation of municipal geographic information systems (GIS). Carnaxide: AUTORIDADE NACIONAL DE PROTECÇÃO CIVIL. 2009. 91 p.

MATTHECK, C. **Updated Field Guide for Visual Tree Assessment.** Karlsruhe: Forschungszentrum Karlsruhe Gm, 2007. 170p.

MATTHECK, C.; BRELOER, H. **The body language of trees:** a handbook for failure analysis. London: The Stationery Office, 1997. 239p.

MENDES, F.H.; SILVA FILHO, D.F.; LOPES, A.M.S. The shadow of buildings as an index of verticalisation in cities. **Hipótese,** Itapetininga, v. 1, n. 1, p. 24-34, 2015.

OKE, T.R. **Boundary layer climates**. 2nd. ed. London: Routledge. 1987. 435p.

OLIVEIRA, J.B.; CAMARGO, M.N.; ROSSI, M.; CALDERANO FILHO, B. **Pedological map of the State of São Paulo:** expanded legend. Campinas: Embrapa Solos / IAC, 1999. 64p.

OLIVEIRA, J.B.C.; PRADO, H. **Semi-detailed pedological map of the State of São Paulo:** Folha Piracicaba (SF 23-Y-A-IV). Campinas, Agronomic Institute of Campinas, 1989. Map, scale 1:100,000.

OLIVEIRA, S.; LOPES, A.M.S. Methodology for assessing the risk of falling trees due to strong winds. The case of Lisbon. In: CONGRESSO DA GEOGRAFIA PORTUGUESA, 6., 2007. Lisbon. **Proceedings...** Lisbon: CGP, 2007. 21 p.

OMM - WORLD METEOROLOGICAL ORGANISATION. **Calculation of monthly and annual 30-year Standard Normals.** WMO. WCDP, n. 10. Geneva, 1989 (Technical document, 341).

PRATA, A.R. **Dimensioning the impact of building height on natural ventilation conditions in the urban environment simulated in a wind tunnel:** the case of Santos. 2005. 243p. Thesis (Doctorate in Architecture and Urbanism) - Faculty of Architecture and Urbanism, University of São Paulo, São Paulo, 2005.

RIBEIRO, P; LOPES, A. Modelling the risk of trees falling on vehicles due to strong winds in Lisbon. In: CONGRESS OF PORTUGUESE GEOGRAPHY, 8., 2011. Lisbon. **Proceedings...** Lisbon: CGP, 2011. p. 1-6.

ROLLO, L.C.P. **Methodologies for quantifying urban green areas:** mapping tree cover and forest inventory of street trees in cities in the State of São Paulo. 2014.103p. Thesis (PhD in Forest Resources) - Escola Superior de Agricultura "Luiz de Queiroz", Universidade de São Paulo, Piracicaba, 2014.

SANTOS, R.M. **Urban morphology and thermal comfort.** AUP 823 Integration Seminar, Faculty of Architecture and Urbanism, 2004. 13p.

SEITZ, **R.A. Manual for pruning forest tree species.** Curitiba: FUPEF, 1995. 56p.

SENE, E.; MOREIRA, J.C. **Geografia geral e do Brasil:** espaço geográfico e globalização. São Paulo: Scipione, 1998. 504p.

SHASHUA-BAR, L.; HOFFMAN, M.E. Vegetation as a climatic component in the design of an urban Street: An empirical model for predicting the cooling effect of urban green areas with trees. **Energy and Buildings**, Lausanne, v. 31, n. 3, p. 221-235, Apr. 2000.

TROEN, I.; PETERSEN E.L. **European Wind Atlas,** Roskild: Riso National Laboratory, 1989. 656p.

CHAPTER 3

ESTIMATION OF LEAF AREA INDEX FOR DIFFERENT URBAN TREE SPECIES

Summary

The scope of this research was to quantify the Leaf Area Index (LAI) of different tree species used in urban afforestation during the dry season in Piracicaba (SP), whose predominant vegetation is stationary semi-deciduous forest. Using the LAI 2000 *Plant Canopy Analyser*, developed by LI-COR, average LAI values were obtained for 15 species. Among the analyses, *Licania tomentosa* (Benth.) Fritsch. obtained the highest value (5.2), followed by *Cupressus sempervirens* L. (5.1) and *Caesalpinia echinata* Lam. (4.3), while the lowest was recorded by *Jacaranda mimosifolia* D.Don (0.8). Hourly estimates of LAI throughout the day generally showed that direct radiation underestimated the Leaf Area Index. As this is an unprecedented study, the estimation of LAI for urban trees attracts attention as it is yet another factor to be considered when choosing a species for use on public roads.

Keywords: Urban trees; leaf area index; LAI 2000 *Plant Canopy Analyser*

3.1 Introduction

The leaf area index (LAI), also known as the *LeafArea Index* (LAI), is the ratio between the leaf area of the canopy and its respective projection onto the ground (m^2 m^{-2}), and is represented by a dimensionless number (WATSON, 1947). LAI is a biophysical variable that is directly linked to ecological and physiological processes, such as photosynthesis and evapotranspiration, as well as helping to understand the physical and chemical adaptations of vegetation (NORMAN; CAMPBELL, 1989).

It can be quantified by direct methods, which, despite being highly accurate, are destructive and labour-intensive, or by indirect methods, which are more practical and more widely used because they can be measured beforehand using the destructive method.

Indirect means include the LAI 2000 *Plant Canopy Analyser, a* device that stands out for its practicality and speed in obtaining the LAI. It analyses the reduction of solar radiation by the canopy, based on the concept of *gap fraction*, which corresponds to the proportion of canopy openings above the measurement point, varying between trees due to the geometric arrangement of the leaves. However, branches and trunks, as well as the clustering of leaves, appear to be the main sources of error because, while the former intercept the radiation, overestimating the result mainly in low LAI locations, the latter damage the assumption of the random distribution of leaves projected onto the plane (WELLES, 1990).

According to Nowak (1994), air quality in cities can be assessed by leaf area, through measurements of pollutant interception for different tree species.

Shinzato et al. (2013) used LAI 2000 inside Trianon Park in São Paulo (SP) for 22 different points and found an average value of 2.52, with estimates ranging from 1.46 to 3.86.

In Portugal, Amorim et al. (2013) studied the influence of ventilation on the dispersion of pollutants and found an overall reduction of 16 per cent in the concentration of CO in an avenue laid out almost parallel to the wind direction when compared to another positioned at approximately 45°. These authors reported the strong dependence of air quality on the synergies between meteorological conditions, 3D configurations of urban *canyons* and the presence of vegetation.

Therefore, this study sought to quantify the leaf area index for trees used in urban afforestation in Piracicaba (SP), a value that is taken into account when calculating the energy balance in urban areas.

3.2 Materials and Methods

The research was carried out at ESALQ/USP, in Piracicaba (SP). Situated at an altitude of 546 m, with geographical coordinates of 22°42'30" S and 47°38'00" W, the municipality has a Cwa climate, according to the Kõppen classification, characterised as humid subtropical with winter droughts; the average annual rainfall is 1328 mm, with average minimum annual temperatures of 14.8°C and average maximum of 28.2°C, according to data from the Centre for Meteorological and Climatic Research Applied to

Agriculture (CEPAGRI) at the University of Campinas (UNICAMP).

Folha de Piracicaba points out that the predominant vegetation in the municipality is stationary semi-deciduous forest, located in the Paulista Peripheral Depression domain (OLIVEIRA; PRADO, 1989).

The LAI 2000 (Figure 3.1) was used to estimate the average LAI for 15 tree species. Three individuals with different crown geometries were considered for each species in order to quantify the standard deviation as well as the average LAI. The field day chosen was 18 July 2014, when the sky was very cloudy and direct solar radiation was practically non-existent, an ideal condition for using LAI, as well as in the morning or late afternoon. According to the Climatological Data Series of the Luiz de Queiroz Campus in Piracicaba, SP (ENGENHARIA DE BIOSSISTEMAS, 2014), the average incident solar radiation for that day was 104.30 W nr^{-2} . In the few moments of direct radiation, there were no LAI measurements. For each tree, one external measurement and five under the canopy were taken, all with the smallest protective cover available (10°), in order to limit the leaf portion of the canopy, avoiding branches and twigs.

Figure 3.1 - LAI 2000 *Plant Canopy Analyser*, next to it a 30 cm ruler and the

user manual. Source: the author

Likewise, in order to assess the variation in LAI values throughout the day under the influence of direct radiation, which is presented as a possible source of estimation error, given the ideal field conditions, 30 July 2014 was used, a day with open skies and little cloudiness. According to the Climatological Data Series of the Luiz de Queiroz Campus in Piracicaba, SP (ENGENHARIA DE BIOSSISTEMAS, 2014), the average incident solar radiation was 234.03 w m^{-2} , with sunrise at 6.45am and sunset at 5.48pm, according to the Centre for Weather Forecasting and Climate Studies of the National Institute for Space Research (CPTEC/INPE, 2014). This was done for the species with the highest and lowest LAI among those studied.

The equipment consists of an optical sensor coupled to a 148° lens that simultaneously measures diffuse radiation in five different angular bands (7^{Q} , *23*^{Q} *, 38-, 53-, 68*^{Q}), with 5 active surface photodiodes arranged in concentric rings (Figure 3.2). The image of the canopy, which is almost hemispherical in shape, is projected onto these rings, allowing the radiation to be measured for each angular interval in the known band and zenith angle, in which an optical filter restricts radiative transmission to wavelengths shorter than 490 nm (LI-COR, 1992). The data collected is stored in the device's internal memory.

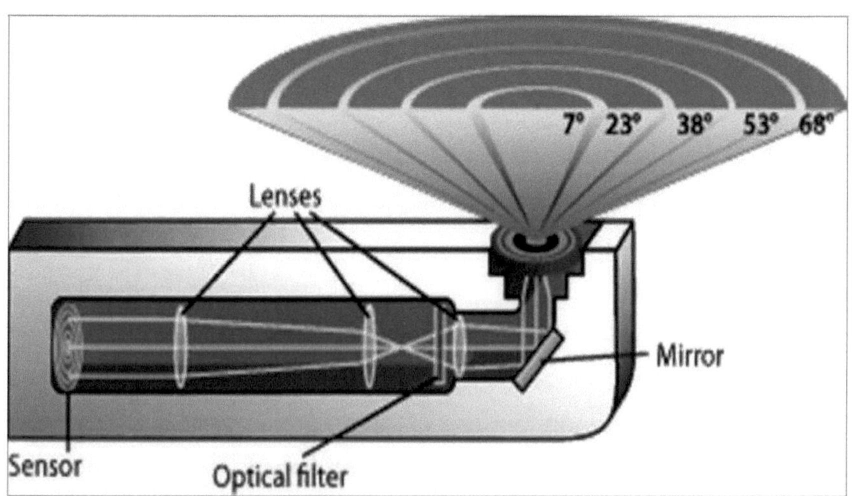

Figure 3.2 - Internal structure of LAI 2000. Source: LI-COR

Then, using the FV 2200 *software* developed by LICOR, the LAI values were downloaded. It is important to note that certain rings can be "masked", for example, if the reference field of view is not large enough, the readings of the last rings need to be disregarded, avoiding the open sky. Figure 3.3 exemplifies this:

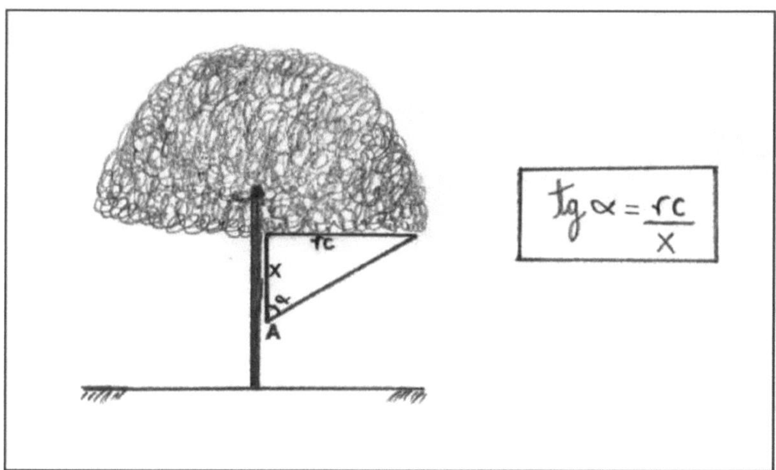

Figure 3.3 - Example of the calculation to be made to accept or exclude the reading, where: A = reading point of the LAI 2000 equipment; a = measurement angle formed between the axis perpendicular to the ground and the end of the canopy (in degrees); rc = radius of the canopy (m); x = distance between the

LAI 2000 equipment and the base of the canopy (m). Source: the author

Thus, once the values of rc and x were known, it was possible to find the angle a, which is the maximum value that can be used. There may be cases where it is necessary to exclude more than one ring. If α> 68°, then all the LAI 2000 readings were under the canopy.

3.3Results and Discussion

Of the 15 species analysed, oiti *(Licania tomentosa* (Benth.) Fritsch.) had the highest LAI, while jacaranda mimoso *(Jacaranda mimosifolia* D.Don) had the lowest. The Italian cypress *(Cupressus sempervirens* L.), despite not being planted in urban areas, had its LAI estimated in order to have a parameter of what a tree with a high LAI would be, as shown in the figure below:

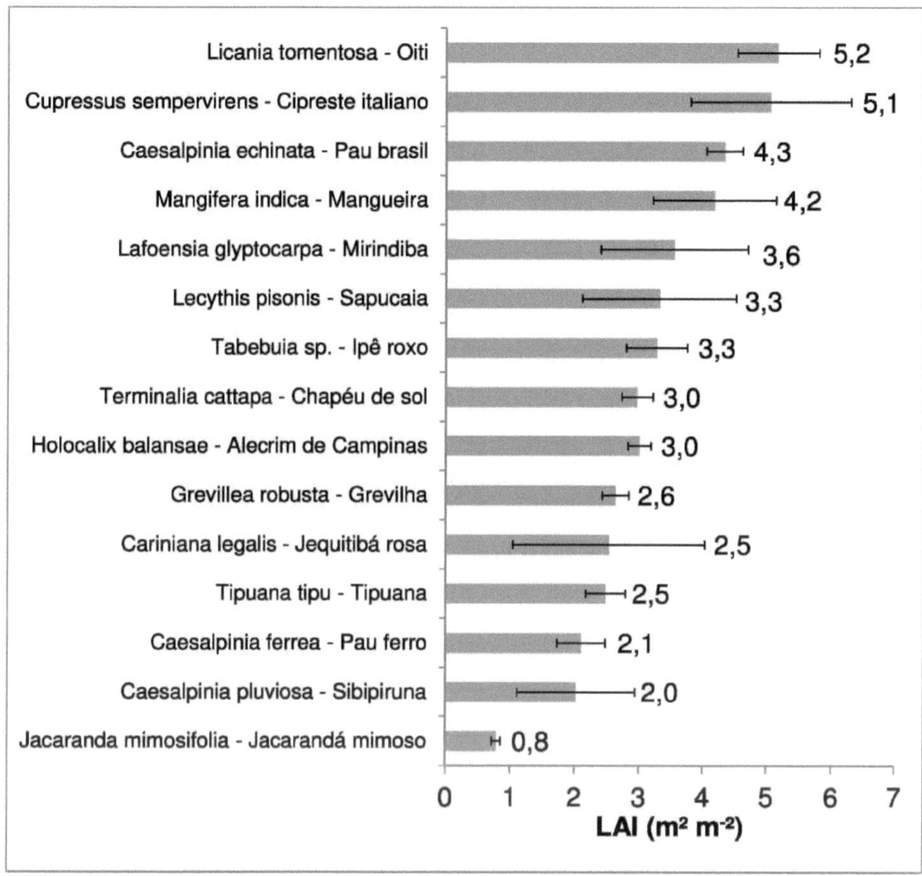

Figure 3.4 - LAI estimates for 15 tree species, with the respective standard deviation. Source: the author

There is a big difference between the LAI of the jacarandá mimoso and the others. The explanation is that, in addition to the fact that the tree does not have a dense canopy, the data was collected in mid-July, corresponding to the local winter and, as it is a deciduous to semi-deciduous species, it had few leaves at the time of quantification, although figure 3.4 shows other species that will still lose leaves in winter. Below you can see the trees with different leaf area indices:

Figure 3.5 - Comparison between (A) *Licania tomentosa* - LAI = 5.2 and (B) *Jacaranda mimosifolia* - LAI = 0.8, respectively. Source: the author

Next, to check the influence of direct radiation on LAI estimates, two species were monitored from 6am until 6pm: *Licania tomentosa*, with high LAI and *Jacaranda mimosifolia,* with low LAI.

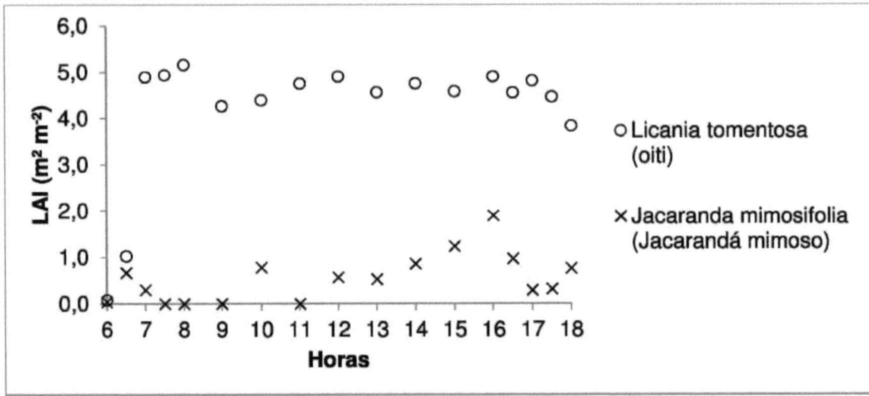

Figure 3.6 - Hourly variation in LAI throughout the day for species with different LAI values. Source: the author

It was clear that the presence of direct radiation influenced the LAI estimate, underestimating the values. The first points for the two species can be considered discrepant due to the time of sunrise, i.e. it was still very dark during the measurement. The second point collected for *Licania tomentosa* was also underestimated due to the presence of a lamppost that was lit at the time of analysis.

3.4 Conclusion

The highest LAI values were found in the species *Licania tomentosa* (Benth.) Fritsch., *Cupressus sempervirens* L. and *Caesalpinia echinata* Lam., while the lowest were found in *Jacaranda mimosifolia* D.Don, *Caesalpinia pluviosa* DC. and *Caesalpinia ferrea* C.Mart.

Hourly monitoring of trees with different LAI estimates showed that direct radiation underestimated LAI. The angle of inclination of the sun may also have interfered with the dispersion curve of the points.

The study of LAI for urban trees deserves more attention. There is little literature on the subject and future research could assess the influence of trees with different LAI on environmental comfort, pollutant retention, rainfall interception and evapotranspiration balance. In view of the above, LAI could be one of the criteria adopted when choosing the species to be used in the afforestation of public roads, in addition to those already mentioned by Paiva (2000), such as type of crown, size, flowers, fruit, trunks, roots, toxicity problems, hardiness, resistance, natural pruning and origin. It is also necessary to develop a scale that groups together trees with similar LAI values.

Adding to the research by Amorim et al. (2013), based on the dispersion of pollutants by winds, for streets with high vehicle traffic, if the prevailing wind is parallel to the street, species with high LAI values can be chosen, while if the prevailing wind is almost perpendicular to the direction of the street, trees with intermediate LAI values may be more interesting.

As the predominant local vegetation is stationary semi-deciduous forest, the data presented here can be compared with measurements in the summer, the rainy season, when the leaves are more vigorous and the LAI tends to be higher.

Due to the existence of three-dimensional computer models, such as ENVI-met, capable of simulating interactions between surface - vegetation - atmosphere, LAI is one of the input variables in the model (BRUSE, 2009), making it possible to insert this information into the plant database. Thus, these trees can be modelled in future microclimate simulations for the site.

References

AMORIM, J.H.; RODRIGUES, V.; TAVARES, R.; VALENTE, J.; BORREGO, C. CFD modelling of the aerodynamic effect of trees on urban air pollution dispersion. **Science of the Total Environment,** Amsterdam, v. 461-462, p. 541-551, Sept. 2013.

BRUSE, M. **ENVI-met 3.1:** Online Manual. 2009. Available at <http://www.envi-met.com/>. Accessed on: 23 July 2014.

CENTRE FOR METEOROLOGICAL AND CLIMATIC RESEARCH APPLIED TO AGRICULTURE. Available at: <http://www.cpa.unicamp.br/outras-informacoes/clima_muni_436.html>. Accessed on: 22 July 2014.

WEATHER FORECASTING AND CLIMATE STUDIES CENTRE / NATIONAL SPACE RESEARCH INSTITUTE. Available at: <http://www.cptec.inpe.br/cidades/tempo/4009>. Accessed on: 30 July 2014.

BIOSYSTEMS ENGINEERING. Meteorological station. Available at: <http://www.leb.esalq.usp.br/posto.html>. Accessed on: 23 July 2014.

LI-COR. **LAI-2000 Plant Canopy Analyser:** Instruction manual. Li-cor Inc. Lincoln, 1992. 182p.

NORMAN, J.M.; CAMPBELL, G.S. Canopy structure. In: PEARCY, R. W.; EHLERINGER, J., MOONEY, H. A., RUNDEL, P. W. (Ed.). **Plant physiological ecology:** field methods and instrumentation. London: Chapman and Hall, 1989. p. 301-326.

NOWAK, D.J. Air pollution removal by Chicago's urban Forest. In: McPHERSON, E.G.; NOWAK, D.J.; ROWNTREE, R.A. (Ed.). **Chicagos Urban Forest Ecosystem:** Results of the Chicago Urban Forest Climate Project. Radnor: USDA Forest Service, 1994. p. 63-81.

OLIVEIRA, PRADO, H. **Semi-detailed pedological map of the State of São Paulo:** Folha Piracicaba (SF 23-Y-A-IV). Campinas, Agronomic Institute of Campinas, 1989. Map, scale 1:100,000.

PAIVA, H.N. Selection of species for urban afforestation. **Revista Ação Ambiental,** Viçosa, v. 2, n. 9, p. 14-16, 2000.

SHINZATO, P.; DUARTE, D.H.S.; BARROS, F.; MOREIRA, D. The impact of vegetation on urban microclimates: estimation of leaf area index in different tree groups. In: ENCONTRO NACIONAL DE CONFORTO NO AMBIENTE

CONSTRUÍDO, 12.; ENCONTRO LATINOAMERICANO DE CONFORTO NO AMBIENTE CONSTRUÍDO, 8., 2013, Brasília. **Proceedings...** Brasilia: ENCAC; ELACAC, 2013.

WATSON, D.J. Comparative physiological studies on growth of field crops: I. Variation in net assimilation rate and leaf area between species and varieties, and within and between years. **Annals of Botany**, London, v. 11, p. 41-76, 1947.

WELLES, J.M. Some indirect methods of estimating canopy structure. **Remote Sensing Reviews,** New York, v.5, p. 31-43, 1990.

CHAPTER 4

MICROCLIMATIC WIND SIMULATION IN A RESIDENTIAL NEIGHBOURHOOD IN PIRACICABA/SP

Summary

This study analysed tree falls in the municipality of Piracicaba/SP, using the Fire Brigade database from 2011 to 2014. The spatial assessment showed that two zones predominated: the part of Vila Rezende/Nova Piracicaba, considered to be wooded, and the other, in the São Judas/São Dimas neighbourhoods, with higher altitudes. The microclimate simulation in the residential neighbourhood (Vila Independência) was able to identify changes in wind direction in the vicinity of the largest buildings, however, this was not enough to channel the wind, but they fell due to the initial wind being strong enough to knock them down.

Keywords: ENVI-met; Computer modelling; Fallen trees

4.1 Introduction

Cities have been consolidated in a disorganised way, with no harmony between green spaces, buildings and living beings, where impermeable surfaces are increasingly common and heat islands occurring more intensely in urban centres, causing discomfort and reducing people's quality of life.

The urban climate, defined by Oke (1978), corresponds to the result of urban modifications to the Earth's surface and the consequent interference in local atmospheric conditions. For Duarte and Serra (2003), there is a set of microclimate mosaics in each city, in which each neighbourhood or smaller territorial units can have their own microclimate, due to the constitution of vegetation, buildings and local soils, which will directly affect the temperature and relative humidity of the air and the direction and speed of the wind.

The action of wind in the urban environment depends on the roughness of the area, which is understood as the friction value of the surface, i.e. the presence of buildings capable of significantly altering the movement of air masses (SANTOS, 2004). An example of this is Lisbon, the Portuguese capital, whose north-south orientated streets are responsible for a significant number of tree falls (RIBEIRO; LOPES, 2011), where the wind speed at which trunks and trees begin to fall corresponds to approximately 7 m s^{-1} (OLIVEIRA; LOPES, 2007).

Following the passage of two hurricanes in the USA in 1995, Duryea (1997) concluded that native species are better adapted to local conditions than exotic ones and that this ensures greater stability for them against natural events.

Moser et al. (2010) analysed the occurrence of damaged trees at the Santa Catarina State University (UDESC) after a strong wind event and found *Cupressus lusitanica, Mimosa scabrella* and *Tipuana tipu* to be susceptible to falls.

In Piracicaba/SP, the highest wind speed ever recorded, according to the ESALQ/USP weather station, was on 29 March 2006, with 44 m s^{-1} , the maximum value that the equipment is capable of measuring. In this tornado event, of F1 intensity, according to the Fujita scale of tornado intensity, more than 3% of the city's trees were felled, 500 of them at ESALQ (Luiz de Queiroz College of Agriculture) alone and another 400 at Unimep (Methodist University of Piracicaba), with urban damage of around R$1.5 million (NUNES et al., 2008).

A computer model capable of carrying out simulations at the microclimate level is ENVI-met. Using this programme, many authors have been able to assess human interference in cities: Ozkeresteci et al. (2003) stated that it is possible to predict climate impacts, Yu and Hien (2006) assessed the benefits of urban parks in Singapore by reducing air temperatures, Kallas (2008) found that urban layout contributed to an increase in air temperature in his study area, Teresina/PI, and Masiero and Souza (2012) proved the microclimatic effects arising from proximity to a dam in São José do Rio Preto (SP).

In view of the relevance of studies related to wind channels in the urban environment, the aim of this research was to assess the influence of strong wind in a residential neighbourhood with falling trees, in the municipality of Piracicaba/SP.

4.2 Materials and Methods

4.2.1 Study area

The municipality of Piracicaba/SP is located at an altitude of 546 metres, with geographical coordinates of 22°42'30" S and 47°38'00" W. The climate is of the Cwa type, with humid summers and dry winters and an average annual rainfall of 1328 mm. The predominant vegetation is stationary semi-deciduous forest, located in the domain of the Paulista Peripheral Depression (OLIVEIRA; PRADO, 1989).

4.2.2 Fallen trees

An organised database was obtained from the 16th Fire Department of Piracicaba/SP containing the address and dates/times of tree falls from 03/01/2011 to 24/05/2014 within the municipal urban perimeter, totalling 275 cases, which were spatialised in GIS using the BatchGeo LLC website[4] . The free version allowed up to 250 records to be spatialised at a time, which had to be done in two parts. As not all the falls fell exactly on the correct geographical coordinate, they were checked one by one, making the necessary adjustments.

Some limitations should be highlighted: (i) It was not possible to obtain data from older falls, with records printed on paper and lost over time (today it is computerised); (ii) Falls that posed some risk to people and/or public property were counted; (iii) The database did not distinguish between falling trees and falling branches; (iv) The date/time of the record corresponds to the time the vehicle left (answering the call) and not exactly the same time as the fall; (v) The Department for the Defence of the Environment (SEDEMA) also has data on tree falls (probably more than the Fire Brigade), but without the time of the falls/calls; however, the Department has made efforts to improve this. In events of greater urgency in dealing with occurrences, both organisations join forces for the same cause, offering mutual support.

---

[4] <http://pt.batchgeo.com/>. Accessed on: 12 August 2015.

4.2.3 The ENVI-met *software*

Understanding the interaction between soil - vegetation - atmosphere is complex (BONAN, 2002). Thus, a microclimate model developed by Michael Bruse and team, initially launched in 1998 (BRUSE; FLEER, 1998), is ENVI-met, a program capable of carrying out simulations aimed at understanding the urban microclimate through structured modelling. It is interesting to use ENVI-met in comparison to other simulation *software*, such as Ansys CFX and Ansys Fluent, because while the former considers the root system of plants, the latter consider them as static objects.

Some precautions regarding the use of ENVI-met were highlighted by Carfan et aL (2012), who noted that after 100 m in height, the wind speed tends to decline and there may be thermal inversion. Rosseti (2013) emphasised that wind speed is constant during processing, with no feedback, and Francisco (2012) warned that wind speed below 0.8 m s^{-1} causes instability in the model. However, Jung et aL (2009) found that the programme simulated the winds well in Busan, South Korea, and the results generated were reliable enough to prevent and verify wind incidences, where the vegetation served as a physical barrier and reduced the initial wind speed of 55 m s by between 18 and 27% .-
1

For the computer simulation, version 3.1 was used, which is made up of two files, .in (area modelling) and .cf (configuration file). Available online, the "Tutorial for beginners: ENVI-met *software* version 3.1"[5] provided support. The model input data is shown in table 4.1:

Table 4.1 - Input data for the ENVI-met model version 3.1. To process the data, it is necessary to comply with the required standards regarding unit separators

Input data	Value
Simulation start date (DD.MM.YYYY)	21.07.2013
Simulation start time (HH:MM:SS)	21:00:00

[5] <http://cmq.esalq.usp.br/wiki/lib/exe/fetch.php?media=publico:projetos:envi-met31_tutorial_iniciantes.pdf>. Accessed on: 12 August 2015.

Total simulation time in hours	1.00
Save model every (min)	60
Wind speed at 10 m height (m s)$^{-1}$	35.7
Wind direction (degrees)	135
Roughness zO at reference point	0.1
Initial atmospheric temperature (K)	305.2
Specific humidity at 2500m (g kg)^{1}	9.4
A relative age at 2m (%)	69
Shortwave adjustment factor	0.5
Grid distance (dx, dy and dz) in metres	4.2
Nesting area *(grids)*	4
Coordinates (lat / lon) in degrees	-22.72 ; - 47.63
Size of the simulated model in *grids*	180 x 140x25

Source: the author

The reasons for choosing the variables were:

(i) 21 July 2013 was a day when there was a strong wind, in which the conventional weather station at ESALQ/USP - Piracicaba/SP, at a height of 10 m, recorded 35.7 m s^{-1} and the automatic station, at 2 m, the predominant direction was southeast;

(ii) The start of the simulation should be a time of neutral atmospheric conditions, in which the potential temperature can be considered constant along the profile of the air mixing layer (STULL, 1988), i.e. the beginning of dusk or before sunrise. As the nearest atmospheric sounding (Campo de Marte station - 83779) is made at 00Z and 12Z (21h00 and 9h00, respectively, local time UTC-3), according to the data compiled by the University of Wyoming[6] , the initialisation of the model at 21h00 is restricted;

(iii) A simulation time of 1 hour was set in order to check the behaviour of the area at the time of the wind gust;

(iv) The area's roughness value (0.1) was adopted as the model's standard, typical for open areas with occasional obstacles (WIERINGA et al.,

[6] <http://weather.uwyo.edu/upperair/sounding.html>. Accessed on: 12 August 2015.

2001);

(v) The initial atmospheric temperature was also obtained from the University of Wyoming, with subsequent correction using the Shodor calculator[7] (Figure 4.1). The value was rounded off because it was a complex numerical model, which would take a long time to process;

Potential Temperature:	305.1868	K	Clear Field
Temperature:	12.8796	C	Clear Field
Pressure:	797.1801	mb	Clear Field

Figure 4.1 - Potential temperature correction using the Shodor online calculator, when entering temperature and pressure data (from the University of Wyoming). Source: Adapted from the Shodor calculator

(vi) For specific humidity, an exponential regression curve was made for 2500 metres above sea level, the data from which is required for the model (Figure 4.2);

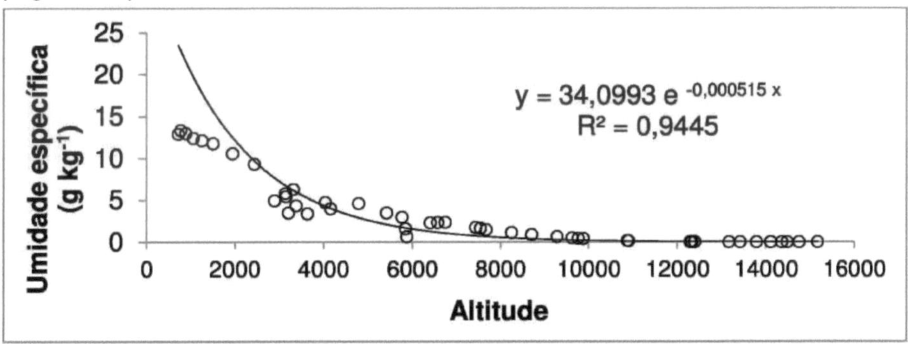

Figure 4.2 - Exponential regression curve to find the specific humidity value at 2500 m altitude, simply by entering 2500 in the x value to find the specific humidity in y (data from the University of Wyoming). Source: the author

(vii) The relative humidity was taken from the automatic station at ESALQ/USP, which is 2 metres above the ground.

(viii) The shortwave adjustment factor was calculated using the balance of the degree 2 polynomial curve of global solar radiation on 21 July 2013

[7] <http://www.shodor.org/os411 /courses/_master/tools/calculators/pottemp/pt1 calc.html>. Accessed on: 12 August 2015.

(Figure 4.3), where the time of 12 noon was taken as the basis for the adjustment of 0.5, which was the closest to reality;

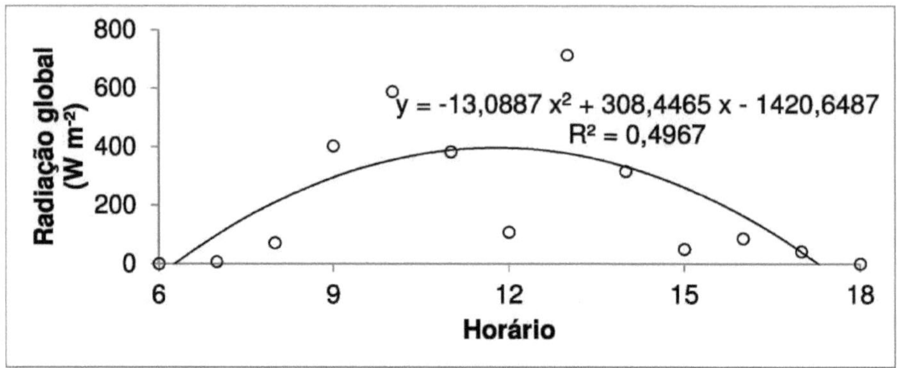

Figure 4.3 - Degree 2 polynomial regression curve to find the global radiation adjustment. In this case, at 12 noon we found 395.93 W m^{-2} Source: the author

(ix) The dx, dy and dz values were 4.20 m. The ENVI-met model can simulate a maximum area of 2 km x 2 km, with a *grid* size of 10 m. The simulated neighbourhood has dimensions of 756 m x 588 m;

(x) The nesting area was 4 *grids*, in which the program itself suggests how much this minimum area should be in order to minimise the effect of the edge by climatic variables.

Other standard ENVI-met variables were used. The vegetation included in the model consisted of 7 different types: grass, sibipiruna *(Poincianella pluviosa* (DC.) L.P. Queiroz), oiti *(Licania tomentosa* (Benth.) Fritsch), false myrtle *(Murraya paniculata* (L.) Jack), palm and two types of shrub (Table 4.2):

Table 4.2 - Vegetation modelling inserted into the model, in the "Plants.dat" file. The LAD *(LeafArea Density)* data was adapted from Mendes et al. (2014), while the RAD *(RootArea Density)* was all used as standard (0.1).

----ENVI-met Database
V Version 3 this file is only usable with ENVI-met
3 ---- V3.0 or higher--------------------------

TT
C T lol- HH. .T LAD LA LA LA LA LA LA LA LA UX
ID? Y m a_fHH T 1 D2 D3 D4 D5 D6 D7 D8 D9 D10 Species

	C		20				0.		0.	0.		0.			0.		grass
gr	3	3	0	0	0.5	0.5	3	0.3	3	3	0.3	3	0.3	0.3	3	0.3	20cm
s	C	1	40	0	10	2	0	0	0	0	0	0.	0.4	0.5	0.	0.3	sibipirun
b	3		0									3			5		a 10m
ot	C	1	40	0	6	2	0	0	0	0.	0.8	0.9	0.9	0.8	0.	0.6	oiti
	3		0							6					7		6m
																	false
f	C		40							0.					0.		myrtle
m	3	1	0	0	5	2	0	0	0	6	0.8	0.9	0.9	0.8	7	0.6	5m
p	C	1	40	0	7	1	0	0	0	0	0	0	0	0	0.8	0.4	palm
m	3		0														tree 7m
a	C	1	40	0	2	1	0.	0.1	0.	0.	0.1	0.	0.1	0.1	0.	0.1	bush
2	3		0				1		1	1		1			1		2m
a	C		40				0.		0.								4m
4	3	1	0	0	4	1	1	0.1	1	0.1	0.1	0.1	0.1	0.1	0.1	0.1	bush

Source: Adapted from Mendes et al. (2014)

4.3Results and Discussion

The tree falls were spatialised in GIS in order to check their dispersion in the urban territory. There was a total correlation with the level of urbanisation in the city, where highly built-up areas (the city centre) had more falls. There were two main concentrations, as shown in figure 4.1. In A, the wooded Nova Piracicaba, Vila Rezende and Jardim Monumento neighbourhoods predominate, while in B, the central area with the highest elevation (approximately 540 m, according to the municipality's contour lines), São Dimas, São Judas, Cidade Jardim, Vila Independência and Vila Monteiro. In this way, a microclimate simulation was carried out for Vila Independência (highlighted in yellow), characterised as residential with approximate dimensions of 756 m x 588 m (Figure 4.4).

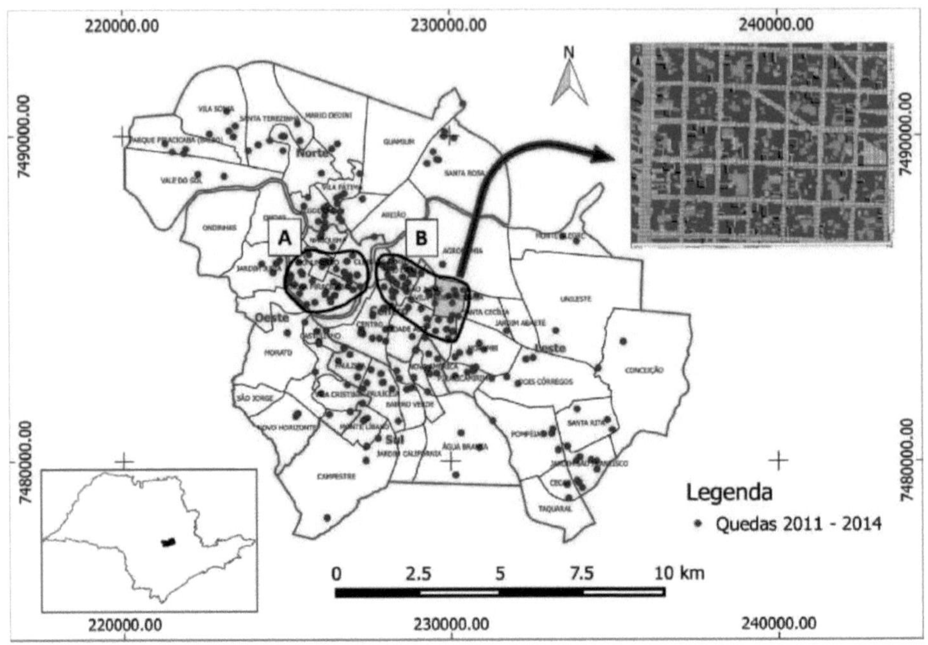

Figure 4.4 - Map of the dispersion of falls in the urban area of Piracicaba/SP (scale 1:100,000), with identification of two regions of greater concentration (A - wooded neighbourhoods and B - neighbourhoods with high relief) and respective modelling of the Vila Independência neighbourhood, with buildings (grey), vegetation (green), asphalt and sandy and clay soils. Source: the author

The heights of the houses and buildings were calculated using *Google Street View,* based on primary wiring at a height of 6 metres, poles at 9 metres, residential houses at 5 metres and each floor of a building at 3 metres. A virtual machine was used in USP's InterNuvem infrastructure environment, with the following configuration: Intel® Xeon CPU E7-2870 @ 2.40 GHz (8 processors in a cluster), with 32 GB of RAM. Longer simulation times with high input wind speeds caused instability in the model and should be avoided when simulating wind gusts. The size of the mesh used (180 x 180 or 250 x 250) does not affect the result, but the simulated area must fit into one of the available models.

Thematic maps were created for the trees that fell in the area, numbered 1 to 6, using the Leonardo 3.75 map viewer, which is a module inserted into

ENVI-met, at a height of 4.2 metres (Figures 4.5 and 4.6), an approximate value to include vegetation whose wind is capable of exerting the moment of force through the arm, promoting leverage and causing the tree to fall.

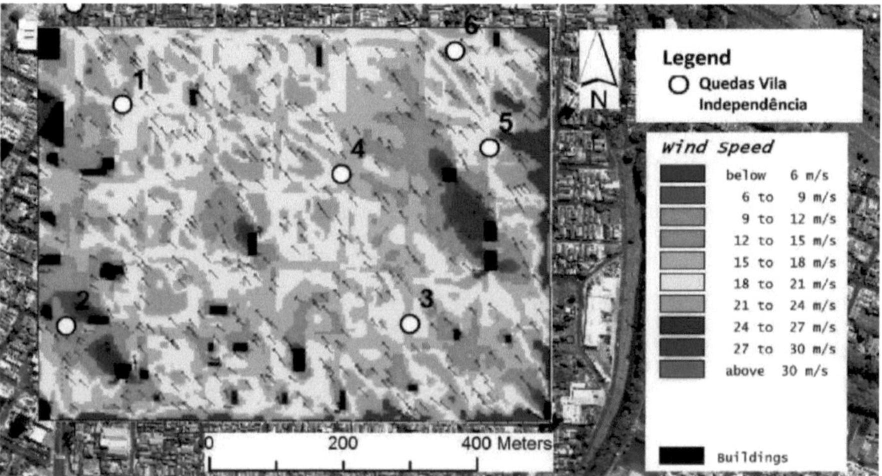

Figure 4.5 - Thematic map of Vila Independência showing the wind speed behaviour in the neighbourhood, with an incoming speed of 35.7 m s^{-1} , facing southeast (135°), at a height of 4.2 m above the ground. Source: the author

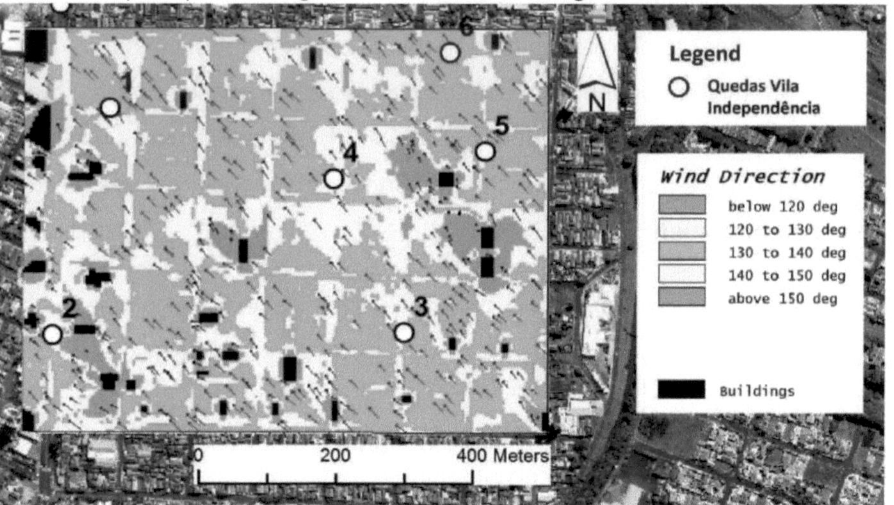

Figure 4.6 - Thematic map of Vila Independência highlighting the change in the initial wind direction of 135° (southeast), at a height of 4.2 m from the ground; in orange, the areas most affected by verticalisation. Source: the author

The average wind speed in the area was 17.6 m s^{-1} , while in the falls it

was 16.6 m s^{-1} , showing no signs of acceleration zones, together with the fact that the average direction in the area was 137° and in the falls 136°. According to Mattheck and Breloer (1997), trees with biomechanical flaws (V-shaped branching, high slenderness coefficient, internal hollow of more than two thirds, cavity at the base and tensioned root) are the most susceptible. It is important to note that it was necessary to exclude the programme's "default" values (wind = 0 m s^{-1} and direction = 180°, when these were over buildings at a height of 4.2 m). It was found that in the vicinity of the buildings that exceeded the cut-off height there were the greatest changes in wind direction, with a deviation of more than 15°, but as there were only a few cases, this did not affect the falls.

4.4 Conclusion

There was no wind channelling for the simulated tree falls, as the Vila Independência neighbourhood has mostly single-storey houses.

The falls occurred because the incoming wind was strong enough to blow them down, where the trees that fell were probably suffering from biomechanical problems or were diseased.

It is important to note that although the ENVI-met model causes instability with high incoming wind speeds simulated for periods of more than 1 or 2 hours, it proved to be an efficient tool for urban planning, as it mapped the behaviour of the neighbourhood based on external conditions, identifying places of greater attention.

References

BONAN, G.G. **Ecological Climatology:** Concepts and Application. New York: CAMBRIDGE UNIV. PRESS, 2002. 990p.

BRUSE, M; FLEER, H. Simulating surface-plant-air interactions inside urban environments with a three dimensional numerical model. **Environmental Modelling & Software,** Oxford, v. 13, p. 373-384, Oct. 1998.

CARFAN, A.C.; GALVANI, E.; NERY, J.T. Study of thermal comfort in the City of São Paulo using ENVI-met model. **Investigaciones Geográficas,** Boletín del Instituto de Geografia, Mexico City, n. 78, p. 34-47, 2012.

DUARTE, D.H.S.; SERRA, G.G. Land use patterns and urban microclimates in the Brazilian continental tropical climate region: correlations and proposal of an indicator. **Ambiente Construído,** Porto Alegre, v. 3, n. 2, p.7-20, apr/jun. 2003.

DURYEA, M. **Wind and Trees:** Surveys of Tree Damage in Florida Panhandle after Hurricanes Erin and Opal. Circular 1183 of the University of Florida Cooperative Extension Service, p. 1-8,1997.

FRANCISCO, R.C.A. **Clima Urbano:** um estudo aplicado a Belo Horizonte, MG. 2012.122p. Dissertation (Master's Degree in Built Environment and Sustainable Heritage) - Federal University of Minas Gerais, Belo Horizonte, 2012.

JUNG, W.S.; PARK, J.K.; LEE, H.W.; KIM, E.B.; CHOI, H.J. Wind speed variation over the leeward region according to vegetation under the strong wind. In: INTERNATIONAL CONFERENCE WIND EFFECTS ON TREES, 2., 2009, Freiburg. **Proceedings...** Freiburg: Univ. Freiburg. Ber. Meteor. Inst., 2009. p. 255-261.

KALLAS, L.M.E. **Designing with climate and vegetation:** a case study of the HBB allotment in Teresina-PI. 2008.167p. Dissertation (Master's in Architecture and Urbanism) - University of Brasília, Brasília, 2008.

MASIERO, E.; SOUZA, L.C.L. Simulation of urban microclimate behaviour. In: SYMPOSIUM OF POST-GRADUATION IN URBAN ENGINEERING, 3., 2012. Maringá. **Proceedings...** Maringá: SIMPGEU, 2012, v. 3, 10p.

MATTHECK, C.; BRELOER, H. **The body language of trees:** a handbook for failure analysis. London: The Stationery Office, 1997. 239p.

MENDES, F.H.; CELLA, A.C.; SILVA FILHO, D.F.; POLIZEL, J.L. Estimation of the Leaf Area Index for different urban tree species. In: CONGRESSO BRASILEIRO DE ARBORIZAÇÃO URBANA, 18., 2014. Rio de Janeiro. **Proceedings...** Rio de Janeiro: CBAU, 2014. 10p.

MOSER, P.; SILVA, A.C.; HIGUCHI, P.; SANTOS, E.M. Post-storm evaluation of the tree planting on the campus of the State University of Santa Catarina, Lages-SC. **Revsbau,** Piracicaba, v. 5, n. 2, p. 40-51.2010.

NUNES, L.H.; CÂNDIDO, D.H.; ARAKI, R.; BARBIN, N.B.C.B.; CASTELLANO, M.S.; COLLAÇO, M.M.; SANTOS, F.R.N.; VICENTE, A.K. Condicionantes físicos e impactos dos tornados do final de março de 2006 no interior paulista. **GEOUSP,** São Paulo, v. 27, p. 99-124, 2008.

OKE, T.R. **Boundary layer climates.** London: Methuen, 1978. 372p.

OLIVEIRA, J.B.C.; PRADO, H. **Semi-detailed pedological map of the State of São Paulo:** Folha Piracicaba (SF 23-Y-A-IV). Campinas, Agronomic Institute of Campinas, 1989. Map, scale 1:100,000.

OLIVEIRA, S.; LOPES, A.M.S. Methodology for assessing the risk of falling trees due to strong winds. The case of Lisbon. In: CONGRESSO DA GEOGRAFIA PORTUGUESA, 6., 2007. Lisbon. **Proceedings...** Lisbon: CGP, 2007. 21 p.

OZKERESTECI, I; CREWE, K.; BRAZEL, A.J; BRUSE, M. Use and evaluation of the ENVI-met model for environmental design and planning: an experiment on linear parks. In: INTERNATIONAL CARTOGRAPHIC

CONFERENCE, 21., 2003, Durban. **Proceedings...** Durban: ICC, 1997, v. 21, p. 402-409.

RIBEIRO, P.; LOPES, A.M.S. Modelling the risk of trees falling on vehicles due to strong winds in Lisbon. In: CONGRESS OF PORTUGUESE GEOGRAPHY, 8., 2011. Lisbon. **Proceedings...** Lisbon: CGP, 2011. 6p.

ROSSETI, K.A.C. **Effects of the use of vegetated roofs on urban heat islands simulated using ENVI-Met software.** 2013. 253p. Thesis (Doctorate in Environmental Physics) - Federal University of Mato Grosso, Cuiabá, 2013.

SANTOS, R.M. **Urban morphology and thermal comfort.** AUP 823 Integration Seminar, Faculty of Architecture and Urbanism, 2004. 13p.

STULL, R.B. **An introduction to boundary layer meteorology.** Dordrecht: Kluwer Academic Publishers, 1988. 666p.

WIERINGA, J.; DAVENPORT, A.G.; GRIMMOND, C.S.B.; OKE, T.R. New revision of Davenport roughness classification. In: EUROPEAN & AFRICAN CONFERENCE ON WIND ENGINEERING, 3., 2001. Eindhoven. **Proceedings...** Eindhoven: EACWE, 2001.8p.

YU, C.; HIEN, W.N. Thermal benefits of city parks. **Energy and Buildings**, Lausanne, v. 38, n. 2, p. 105-120, Feb. 2006.

CHAPTER 5

MICROCLIMATIC CHANGES AT JOSÉ BONIFÁCIO SQUARE IN PIRACICABA (SP)

Summary

The aim of this research was to assess the effects on the urban microclimate of the José Bonifácio Square in Piracicaba (SP) after its refurbishment in 2005. *Computational Fluid Dynamics (*CFD) simulations were carried out of two different situations in the square: before and after the renovation, with images from 2004 and 2013, using ENVI-met 3.1 *software.* The results showed that changes in the order of 25 per cent of the initial vegetation, as well as the replacement of the type of pavement around the square, contributed to a difference of up to 0.5 K in air temperature and a respective 4 per cent decrease in relative humidity in the area where the renovation was most intense.

Keywords: Urban environment; Computer modelling; Microclimate

5.1 Introduction

Urban vegetation plays a significant role in promoting human thermal comfort. Soil sealing and the removal of vegetation cause a series of changes to the microclimate which, by making urban areas warmer, intensify the heat island effect. In view of this, tree canopy cover appears to be an important tool in terms of the thermal comfort and well-being of its inhabitants in open spaces, by attenuating sunlight through the foliage, since it increases the relative humidity of the air and reduces the temperature.

Oke (1978) defines the urban climate as the result of changes resulting from the urbanisation of the Earth's surface and its consequent interference in local atmospheric conditions.

A study carried out in Tel Aviv by Shashua-Bar and Hoffman (2000) showed that the cooling effect in vegetated areas is markedly noticeable in the neighbourhood, up to around 100 m, with 80% of the local cooling contributed by the shade of the trees. In tree-lined streets in the Mediterranean region of Israel, there was a temperature reduction of up to 2.5 K compared to non-wooded streets (SHASHUA-BAR; HOFFMAN, 2004), highlighting the role of vegetation in urban microclimate regulation.

Still on a microclimatic scale, to assess climatic influences resulting from changes in external environments, recent research is being carried out using the ENVI-met *software*. Developed by Bruse and Fleer (1998), the model calculates the energy balance for short and long waves using climate data such as: air temperature (K), relative humidity (%) and specific air humidity ($g\ kg^{-1}$), global radiation ($W\ m^{-2}$), wind direction and speed ($m\ s^{-1}$). As for visualising the results generated by the simulations, during the installation of ENVI-met, it is possible to install the Leonardo *software*, which is a binary file visualiser (extension .EDI and .EDT).

With the advanced ENVI-met and Leonardo numerical models, it has been possible to predict climatic impacts, thus making it possible to design urban parks with greater comfort, with a view to using them appropriately and with minimal environmental impact (OZKERESTECI et al., 2003).

Also using computer technology with ENVI-met, in a parametric study in Hong Kong, Ng et al. (2012) tested the effects of urban form and vegetation on the microclimate and found that adding 16% tree cover to the total area analysed reduced the air temperature by 0.4 K and, with a 56% increase in vegetation, there was a drop of up to 1.8 K.

As the object of study, Praça José Bonifácio represents an important urban and historical landmark for the municipality of Piracicaba. With an area of approximately 17,000 $metres^2$, it was renovated in 1980 when vehicle access to its perimeter was radically restricted. This controversial solution accelerated the process of urban degradation of the square, as it prevented direct access to the surrounding housing and thus limited its period of greatest occupation to business hours. Based on these findings, the *design* was remodelled in 2005, promoting the circulation of vehicles in its surroundings.

That said, the aim of this study was to assess the changes in the square's microclimate resulting from these interventions, by modelling the area for the years 2004 and 2013.

5.2 Materials and Methods 5.2.1 Study area

Praça José Bonifácio is located in the centre of Piracicaba (SP), coordinates 22°43'29" S and 47°38'56" W. Simulations were carried out in ENVI- met version 3.1 were carried out for the following situations: (i) 2004 image, with no asphalted streets on the perimeter and a greater number of trees, especially on the part opposite the Cathedral Church of Santo Antônio and (ii) 2013 image, with asphalted streets on the perimeter and a reduction in the number of trees due to the installation of a taxi rank on the north-eastern stretch, as shown in figure 5.1. Figure 5.2 shows the current situation, with the Cathedral Church of St Anthony in the background.

Figure 5.1 - Images of José Bonifácio Square in 2004 (A) with the infrared band to emphasise the vegetation, taken from an overflight, and in 2013 (B) taken from Google Earth. Source: the author

Figure 5.2 - José Bonifácio Square and its surroundings. Source: IPPLAP

In order to accurately verify the interference of the renovation on the microclimate, the buildings surrounding the square and the model's input data configuration file were kept in both simulations. In this way, once all the variables have been fixed, the results of the simulations correspond exactly to the changes made when the square was remodelled.

5.2.1 The ENVI-met *software*

The ENVI-met *software* is capable of carrying out urban fabric modelling simulations through interactions between surface - vegetation - atmosphere, making it possible to insert vegetation and other constructive elements, altering the morphology of the area to be studied.

The same configuration file was used for the model's computer simulations (Table 5.1).

Table 5.1 - Input data to the *software,* also known as the configuration file. In order to process the data, it is necessary to comply with the required standards regarding unit separators

Input data	Value
Simulation start date (DD.MM.YYYY)	15.07.2004
Simulation start time (HH:MM:SS)	21:00:00
Total simulation time in hours	48.00
Save model every (min)	60
Wind speed at 10 m height $(m\ s)^{-1}$	2.1
Wind direction (degrees)	110

Roughness zO at reference point	0.1
Initial atmospheric temperature (K)	290.6
Specific humidity at 2500m (g kg)$^{-1}$	3.2
Relative humidity at 2m (%)	83
Shortwave adjustment factor	0.5
Grid distance (dx and dy) in metres	4.0
Nesting area *(grids)*	4
Coordinates (lat / lon) in degrees	-22.72 ; -47.65
Size of the simulated model in *grids*	60 x 84 x 20

Source: the author

The ESALQ/USP meteorological station is located 3.3 km away from the study area, from which values were obtained for initial atmospheric temperature, relative humidity, prevailing direction and wind speed (all referring to the average for July 2004). It is important to note that some of the data is available in the Julian calendar and, as 2004 is a leap year, attention should be paid to handling the correct data. As for the specific humidity at 2500 m, the data was extracted from the University of Wyoming to create the scatter plot with the exponential regression curve for the days of July 2004 (Figure 5.3). Other standard ENVI-met variables were used.

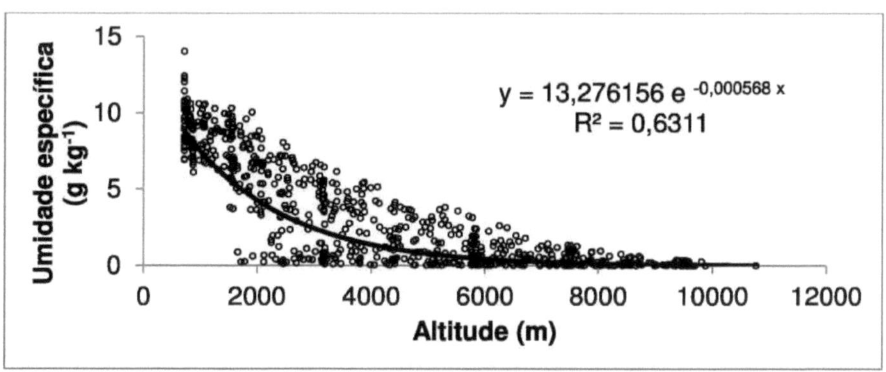

Figure 5.3 - Regression curve to find the value of specific humidity at 2500 m, as one of the input variables in the model. Source: the author

The nesting area corresponds to the transition area around the model and the program itself suggests how much the minimum area should be so that the effect of the model's edge is minimised.

5.2.3Calibration and adjustments to the simulation model

The ENVI-met model was calibrated to the conditions at the study site. The first 24 hours of each simulation were discarded in order to ensure the stability of the programme, according to the methodology adopted by Assis, Sirqueira and Bamberg (2013).

The three-dimensional model consists of a horizontal plane containing the edge area and a vertical profile of the mesoscale at an altitude of 2500 m above sea level, in which there is a single input of data to describe initial conditions in its various layers: soil, surface (including paving, building and vegetation profiles) and atmosphere.

The vegetation included in the model consisted of trees between 10 and 15 metres high, with a dense to very dense canopy. The maps generated show the values found at the average height of pedestrians (1.5 metres).

In the atmosphere layer, the model predicts air flow (speed and direction), turbulence, potential temperature, air humidity and short- and long-wave radiation fluxes, based on the fundamental laws of fluid dynamics and thermodynamics. In the soil layer, the model calculates the vertical profile of air temperature and humidity from the surface, while the hydrological and thermodynamic processes vary according to the properties of each type of soil composition.

The model defines a neutral atmosphere initialisation condition (static stability) when the potential temperature can be considered constant along the profile of the air mixing layer (STULL, 1988). Therefore, it is important to start the simulation in neutral atmosphere conditions, such as early evening or before sunrise. As the nearest atmospheric sounding (Campo de Marte - station number 83779) is carried out at 00Z and 12Z (21h00 and 9h00), given that the square is 3 zones west of the Greenwich meridian, the initialisation of the model at 21h00 is restricted.

5.3 Results and Discussion

The simulations were run on an Intel® Xeon CPU E7- 2870 @ 2.40 GHz processor (8 processors in a cluster), one with 16GB of RAM and the other with 32 GB of RAM. Processing times ranged from 21 to 24 hours, which is considered optimal given the complexity of the ENVI-met models.

Based on the modelling of the area for different situations (Figure 5.4), it was observed that the time with the greatest thermal amplitude between the two situations was at 3pm (Figure 5.5), where the values were extracted using receivers inserted into the models, in the region of the renovation, near the taxis (point 46 x 70 on the *grid).* With approximately 25% modification of the initial tree cover, a difference of up to 0.5 K was found.

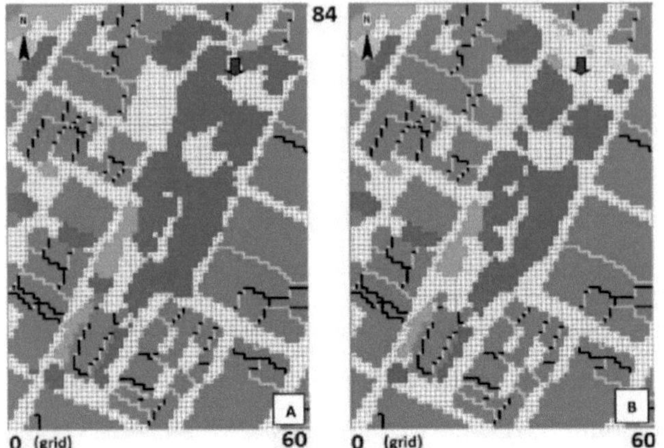

Figure 5.4 - Modelling of the areas in different configurations: 2004 (A) and 2013 (B), together with a receiver installed to analyse climatic variations in the part of the taxis to the north-east of the image, indicated by the arrow. Source: the author

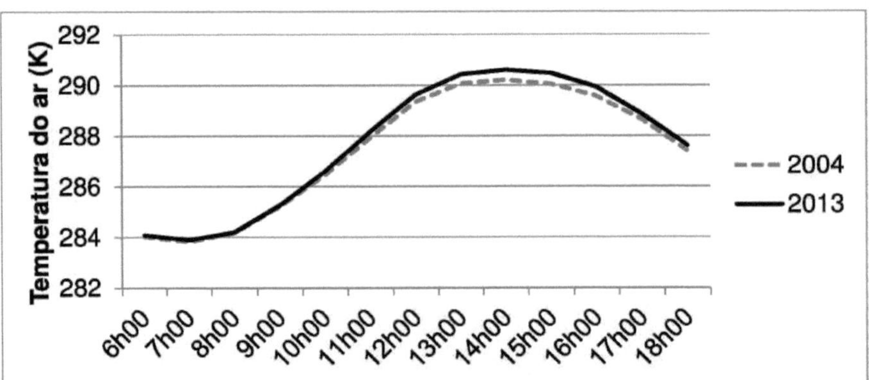

Figure 5.5 - Variation in air temperature in the simulated situations, obtained by the ENVI-met receiver. Source: the author

The results of the simulations can be seen in figure 5.6 below. It shows the spread of the heat spot and the reduction in the relative humidity spot.

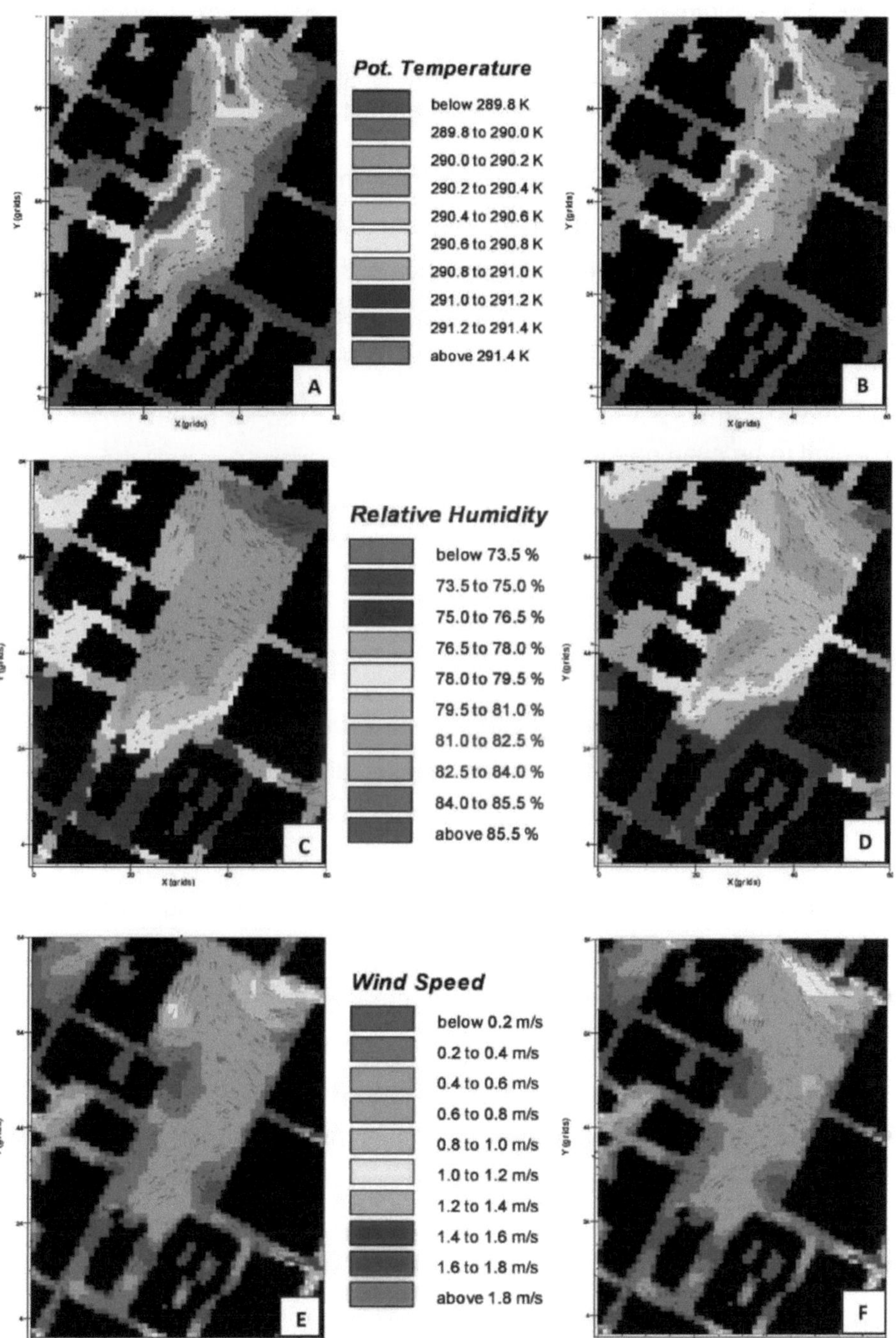

Pot. Temperature
below 289.8 K
289.8 to 290.0 K
290.0 to 290.2 K
290.2 to 290.4 K
290.4 to 290.6 K
290.6 to 290.8 K
290.8 to 291.0 K
291.0 to 291.2 K
291.2 to 291.4 K
above 291.4 K
A
B
Relative Humidity
below 73.5 %
73.5 to 75.0 %
75.0 to 76.5 %
76.5 to 78.0 %
78.0 to 79.5 %
79.5 to 81.0 %
81.0 to 82.5 %
82.5 to 84.0 %
84.0 to 85.5 %
above 85.5 %
C
D
Wind Speed
below 0.2 m/s
0.2 to 0.4 m/s
0.4 to 0.6 m/s
0.6 to 0.8 m/s
0.8 to 1.0 m/s
1.0 to 1.2 m/s
1.2 to 1.4 m/s
1.4 to 1.6 m/s
1.6 to 1.8 m/s
above 1.8 m/s
E
F

Figure 5.6 - Simulation results for the variables air temperature, relative humidity and wind speed, for 2004 and 2013 respectively, at a height of 1.5 metres above ground level. The arrows scattered across the simulation represent the wind direction. Source: the author

The change in the material used for the floor covering around the perimeter of the square (light to medium coloured semi-permeable with asphalt) combined with the removal of some trees increased the amount of heat absorbed, leading to an increase in air temperature and a decrease in relative humidity. The effect can be seen by comparing figures 5.6 (A) and 5.6 (B) or 5.6 (C) and 5.6 (D). In the hottest periods of the year, the amplitude is expected to be even greater.

With regard to the wind, in Figures 5.6 (E) and 5.6 (F), when it originates from the south-east, it is clear from the simulation that there may be an increase in its speed.

5.4 Conclusion

The replacement of the light, semi-permeable flooring (in 2004) with dark, impermeable flooring (in 2013) led to increased absorption of solar radiation and a consequent increase in air temperature of 0.3 K around the perimeter of the square. In the new taxi area (Figure 5.1 B), to the north-east of the image of the square, where most of the trees have been removed, microclimatic changes were also seen due to the renovation, with an increase of up to 0.5 K in air temperature and a respective 4% decrease in relative humidity.

The changes made to the square were necessary to revitalise the public space; however, a solution that also improves the microclimate is possible and feasible through small changes such as adding trees that provide shade over the asphalt and replacing the type of pavement around the perimeter with more permeable material.

Finally, the importance of using the CFD-based simulation programme is highlighted, with a view to improving the quality of urban life by identifying areas that deserve greater attention and future interventions.

References

ASSIS, E.S.; SIRQUEIRA, C.A.; BAMBERG, A.M. Influence of vegetation on microclimate in a simulated controlled environment. In: ENCONTRO NACIONAL DE CONFORTO NO AMBIENTE CONSTRUÍDO, 12.; ENCONTRO LATINOAMERICANO DE CONFORTO NO AMBIENTE CONSTRUÍDO, 8., 2013, Brasília. **Proceedings...** Brasilia: ENCAC; ELACAC, 2013. 10p.

BRUSE, M; FLEER, H. Simulating surface-plant-air interactions inside urban environments with a three dimensional numerical model. **Environmental Modelling & Software,** Oxford, v. 13, p. 373-384, Oct. 1998.

NG, E.; CHEN, L; WANG, Y.; YUAN, C. A study on the cooling effects of greening in a high-density city: An experience from Hong Kong. **Building and Environment,** Oxford, v. 47, p. 256-271, Jan. 2012.

OKE, T.R. **Boundary layer climates.** London: Methuen, 1978. 372p.

OZKERESTECI, I; CREWE, K.; BRAZEL, A.J; BRUSE, M. Use and evaluation of the ENVI-met model for environmental design and planning: an experiment on linear parks. In: INTERNATIONAL CARTOGRAPHIC CONFERENCE, 21., 2003, Durban. **Proceedings...** Durban: ICC, 1997, v. 21, p. 402-409.

SHASHUA-BAR, L.; HOFFMAN, M.E. Vegetation as a climatic component in the design of an urban Street: An empirical model for predicting the cooling effect of urban green areas with trees. **Energy and Buildings**, Lausanne, v. 31, n. 3, p. 221-235, Apr. 2000.

SHASHUA-BAR, L.; HOFFMAN, M.E. Quantitative evaluation of passive cooling of the UCL microclimate in hot regions in summer, case study: urban streets and courtyards with trees. **Building and Environment,** Oxford, v. 39, p. 1087-1099, Sept. 2004.

STULL, R.B. **An introduction to boundary layer meteorology.** Dordrecht: Kluwer Academic Publishers, 1988. 666p.

CHAPTER 6

FINAL CONSIDERATIONS

Based on the results obtained in this book, he concludes:

The behaviour of falling trees in Piracicaba/SP has been shown to be affected by urbanisation. Densely built-up areas, such as the urban centre, with stressful conditions for tree growth, compacted soil and possible cuts and damage to the roots directly affected support.

Most of the falls (78.0 per cent) occurred during spring and summer, seasons linked to wind and rain, in that order. Therefore, the greatest care should be taken during this period.

The predominance of winds from the fourth quadrant, with easterly, south-easterly and southerly directions, suggests that the most suitable locations for industrial parks, which emit polluting gases, are on the opposite side, i.e. the second quadrant, with northerly, north-westerly and westerly directions, so that these gases do not enter the city and harm people's health.

The species most susceptible to falling were: *Pachira aquatica* Aubl. (monguba), *Handroanthus* sp. (ipê roxo) and *Tipuana tipu* (Benth.) Kuntze (tipuana). The monguba, in addition to its dense crown, which increases the surface area in contact with the wind and suffers leverage, has vigorous roots, which spread over the tarmac and pavement, whose pruning in this area compromises support.

The ENVI-met v. 3.1 computer program identified zones capable of modifying wind speed and direction, with a good representation of the roughness of the area analysed.

In practice, it is very difficult to stop trees falling. Therefore, on windy and rainy days, you should avoid standing under trees, as well as driving or parking cars in the area of influence. As this is a hot city (humid subtropical climate with winter droughts and relief located in the Paulista Peripheral Depression), where shade is highly valued, care should be taken to ensure that vegetation remains.

The application of the methodology developed proved to be satisfactory and can be used in other studies, in which the complementation with more data and the age of the trees will certainly add more quality to the subject of tree falls on public roads.

Printed by Books on Demand GmbH, Norderstedt / Germany